SALT
Your Way to Health

2nd Edition

D1530593

David Brownstein, M.D.

For further copies of *Salt Your Way To Health,2nd Edition*

Call: **1-888-647-5616** or send a check or money order in the amount of: $20.00 ($15.00 plus $5.00 shipping and handling) or for Michigan residents $20.90 ($15.00 plus $5.00 shipping and handling, plus $.90 sales tax) to:

Medical Alternatives Press
4173 Fieldbrook
West Bloomfield, Michigan 48323

ISBN: 978-0-9660882-4-3
Medical Alternatives Press
4173 Fieldbrook
West Bloomfield, Michigan 48323
(248) 851-3372
(888) 647-5616
www.drbrownstein.com

Acknowledgements

I gratefully acknowledge the help I have received from my friends and colleagues in putting this book together. This book could not have been published without help from the editors— my wife Allison and my chief editor Janet Darnell. Special thanks to Angela Doljevic for all her help with the cover. Thank you to Jessi Brownstein for her help with the title pages.

I would also like to thank my patients. It is your search for safe and effective natural treatments that is the driving force behind holistic medicine. You have accompanied me down this path and I appreciate each and every one of you.

And, finally, I would like to thank my staff. From the bottom of my heart THANK YOU! Thank you for believing in me and believing in what we do. Without your help, this would not be possible.

A Word of Caution to the Reader

The information presented in this book is based on the training and professional experience of the author. The treatments recommended in this book should not be undertaken without first consulting a physician. Proper laboratory and clinical monitoring is essential to achieving the goals of finding safe and natural treatments. This book was written for informational and educational purposes only. It is not intended to be used as medical advice.

ABOUT THE AUTHOR

David Brownstein, M.D. is a family physician who utilizes the best of conventional and alternative therapies. He is the Medical Director for the Center for Holistic Medicine in West Bloomfield, MI. Dr. Brownstein is a graduate of the University of Michigan and Wayne State University School of Medicine. He is Board-Certified by The American Academy of Family Practice. Dr. Brownstein is a member of the American Academy of Family Physicians and the American College for the Advancement in Medicine. He is the father of two wonderful girls, Hailey and Jessica. Dr. Brownstein has lectured internationally about his success using natural items. Dr. Brownstein has authored nine books: *Drugs that Don't Work and Natural Therapies That Do, 2nd Edition, The Miracle of Natural Hormones 3rd Edition, Overcoming Thyroid Disorders 2nd Edition, Overcoming Arthritis, Iodine: Why You Need It, Why You Can't Live Without It 4th Edition, The Guide to Healthy Eating 2nd Edition, Salt Your Way to Health 2nd Edition, The Guide to a Gluten-Free Diet, and The Guide to a Dairy-Free Diet.*

Dr. Brownstein's office is located at:

Center for Holistic Medicine
5821 W. Maple Rd.
Ste. 192
West Bloomfield, MI 48323
248.851.1600
www.drbrownstein.com

Dedication

To the women of my life: Allison, Hailey and Jessica, with all my love.

And, to my patients. Thank you for being interested in what I am interested in.

Contents

Preface to First Edition

For years, we have heard the following: A low-salt diet is healthy. There is no difference between table salt and sea salt. Low-salt products are better for you.

All of these comments are believed to be true, both by physicians and lay people. This book was written to provide a different viewpoint for the reader. _Salt Your Way to Health_ will challenge each of the above statements and give you a healthier alternative to regular table salt.

I became interested in the therapeutic uses of salt ten years ago after observing that many of my patients were doing poorly on a low-salt diet. Most patients have a difficult time staying on such a diet because of the dull taste of low-salt food. In addition, I saw little clinical benefit from a low-salt diet.

In my search for safe and natural holistic remedies I came across unrefined salt. Unrefined salt contains over 80 essential minerals. Refined salt has no minerals. From the beginning of my holistic practice, I found most patients deficient in minerals. Upon learning of the mineral content of unrefined salt, it became clear that unrefined salt would be a much healthier choice than refined salt.

I was leery about using salt, especially in those patients with hypertension. I was taught in medical school that salt=hypertension and that you could lower blood pressure by putting people on a low-salt diet. However, my clinical experience has shown that many patients with hypertension had a significant improvement in their blood pressure when mineral deficits were corrected with mineral supplementation. When I began using unrefined salt as part of a holistic regimen to correct mineral deficits, a funny thing happened. My patients with high blood pressure began improving. Their blood pressure began to fall. In addition, I found that unrefined salt helped improve the immune system and the hormonal system, as well as other areas of the body.

How could salt be helping all these conditions when I had been taught that salt was bad? I began to search the medical literature and I was struck by what I found.

The literature showed that the connection between salt and hypertension was very weak at its best. In fact, my clinical experience showed there was little benefit in restricting salt in order to control blood pressure. This book will explore this concept in much more detail.

I have seen patients with fibromyalgia, chronic fatigue, dizziness, Meniere's disease and other health problems improve when they begin using unrefined salt in conjunction with a holistic regimen.

This book was written to give the reader a new look at salt. Unrefined salt is a vital substance that is a wonderful source of minerals. It helps support the thyroid and adrenal glands. My clinical experience has shown that it is impossible to have an optimally functioning immune system when there is a salt-deficient state present.

I hope this book helps you discover a healthier lifestyle.

To all of our health!!

David Brownstein, M.D.

January, 2005

Preface to Second Edition

It has been five years since I wrote the first edition of ***Salt Your Way to Health.*** Since that time I have treated thousands of patients with unrefined salt. I continue to support the use of unrefined salt for promoting health. In fact, time has only assured me of the benefits of using unrefined salt in any dietary regimen.

Recently an article appeared on Reuters that was titled, "Cutting back on salt could save U.S. billions." The article stated that, "Reducing daily intake (of salt) to 1,500mg may save $26 billion in care costs."[i] Most Americans eat more salt than that—around 3,500mg/day. Do we need to eat less salt?

The answer to that question depends on what type of salt we are talking about. If we are talking about refined salt, the answer is yes, we do need to eat less salt. In fact, we don't need to eat any refined salt as it is a devitalized food source. Refined salt does not contain minerals and its use leads to the depletion of minerals in the body. The use of refined salt is certainly harmful and needs to be avoided.

If the question was referring to unrefined salt, then the answer is no, we need to eat more unrefined salt. Unrefined salt is a vital substance our bodies require in order to optimally function.

I believe a more reasonable headline for the Reuter's article could state, "Cutting out refined salt and replacing it with unrefined salt could save U.S. billions and improve health."

I hope this book provides you with the information you need to incorporate the use of unrefined salt in your diet. I have found it nearly impossible to help my patients overcome illness and achieve their optimal health if they do not ingest unrefined salt in their diet. It amazes me how many times I see laboratory reports indicate patients are actually salt-deficient.

I write my books to educate you about how to improve your health. I believe that once you read about the benefits of unrefined salt, you will come to the same conclusion I have come to.

To All of Our Health,

David Brownstein, M.D.

January, 2010

[i] Reuters. 9.9.2009

Chapter 1

The History of Salt

History of Salt

Introduction

Without salt, life itself would not be possible. Salt is as important to life as oxygen and water. All living things utilize salt as an essential ingredient for life. In ancient times, salt was as valued a commodity as gold. The various uses of salt are referred to in the history of many cultures, including the ancient Hebrews, Chinese, Greeks, and Romans.

The earliest known writing about salt occurred about 5,000 years ago in China. And, nearly 3,500 years ago, the ancient Egyptians recorded pictures of salt production.

Salt was one of the first commodities known to mankind. Salt bars were used as a currency for more than 1,000 years in

Ethiopia. In fact, until a little over 150 years ago, salt was still one of the most used commodities worldwide.

In past times, governments were financed on salt taxes. The first to tax salt was a Chinese emperor Hsia Yu (2,200 BC). In ancient Greece, salt was exchanged for slaves. In ancient Rome, soldiers were paid in "solarium argentums" which became the English word "salary".[1] Wars were not only fought over, they were paid for with salt. Throughout history, salt was at the forefront of wars, revolutions, and governmental policies.

Salt was an important item for many of the ancient religions. It is the symbol to the Jews of God's covenant with Israel. In the Torah, it is written, "The Lord God of Israel gave the kingdom over Israel to David forever, even to him, and to his sons, by a covenant of salt." Salt was thought to protect against evil in Islam and Judaism. There are 32 references to salt in the Bible such as "salt of the earth." Christianity recognizes salt as being associated with longevity.

Hippocrates (450 BC) recognized the healing powers of salt. He recommended many salt remedies for healing infections, clearing congestion and curing diseases of the spleen. Paracelsus (1493-1541 A.D.) wrote, "The human being must have salt, he cannot be without salt. Where there is no salt, nothing will remain, but everything will tend to rot."

Before refrigeration, salt was invaluable as a preserving agent. Countries could trade food with each because of the ability of salt to preserve the food for its journey. Meat, fish, butter and other foods were preserved with salt prior to being shipped. Before refrigeration, societies could not survive unless they had adequate amounts of salt to preserve their food. In modern times, salt is still used as a preservative in many foods such as fish and meat.

In the American Civil War, the North was successful in limiting the availability of salt to the South. "Salt is eminently contraband, because {of} its use in curing meats, without which armies cannot be subsisted."[2] Part of the Union strategy was to disrupt the South's supply of salt, which it did effectively. Throughout the Civil War, the Union attacked the South's salt production facilities.

The South's position became more tenuous as their salt supplies dwindled. They could not preserve their food, which meant their army could not be properly fed. There is no doubt that a decline in the availability of salt played a large role in the fall of the South.

As Americans began moving west after the Civil War, their movements were tied to finding sources of salt. Salt was necessary to process silver and other minerals. Wherever

Americans went in the Western United States, salt mining and production followed.

Many large cities actually sit atop salt mines. Detroit and Cincinnati are examples of two such cities.

The British were financing their colonization of India via a salt tax. In the 1930's, Mahatma Gandhi organized a protest of this high salt tax.

In the twentieth century, the uses of salt for manufacturing purposes multiplied exponentially including its use in the manufacturing of pulp and paper, soap, dyes for color, and for its use in clearing snow and ice from the roadways. Nuclear waste products are stored in salt mines. Food is still preserved in salt.

In medicine, salt is used as part of an intravenous solution. In the emergency room, IV's of "normal saline" are an invaluable treatment in the acute care setting.

Final Thoughts

Salt is an integral part of our modern world. This book will explore many of the health benefits of unrefined salt and show you why you need to avoid refined salt.

---

[1] Salt Institute. From saltinstitute.org
[2] www.geology.ucdavis.edu

Chapter 2

The Difference Between Refined and Unrefined Salt

The Difference Between Refined and Unrefined Salt

Introduction

Salt is salt, right? Of course not. Just as there are many different brands of cars, there are many differences in various salt brands. This chapter will focus on the difference between refined and unrefined salt.

Salt in its natural form is referred to as unrefined salt. Unrefined salt has not been altered by man. Therefore, it contains many different minerals and elements that are useful for the body. For example, unrefined sea salt (e.g., Celtic Sea Salt) contains over 80 minerals and elements—all the natural elements necessary for life. This is contrasted with refined salt which contains two major items: sodium and chloride.

Refined Salt

How is salt refined? Most commercial refined salt has been harvested mechanically from various salt mines as brine. Brine is a highly concentrated solution of water and salt. Prior to mechanical evaporation, the brine is often treated with chemicals to remove minerals (which are sold for use in industry). The minerals are referred to as "impurities" in salt. These chemicals used to treat refined salt can include sulfuric acid and/or chlorine. Next, water is evaporated under high compression and heat which disrupts the molecular structure of salt. Finally, almost all of the remaining moisture in the salt is removed in a fluidized-bed dryer.[1]

All food-grade salt available in the U.S. must comply with the National Academy of Science's Food Chemicals Codex Sodium Chloride Monograph (1996). Up to 2% of food-grade salt may contain anti-caking, free flowing, or conditioning agents. These agents may include sodium ferrocyanide, ammonium citrate, and aluminum silicate. None of these products have any positive effects in the body. Dextrose, also known as refined sugar, is used as a stabilizer so that iodide will stay in the salt.

The final purity of food-grade salt is between 99.7-99.95% "pure". Pure refers to the sodium and chloride content. The other "impurities", including healthy minerals and elements, have been removed from refined salt. Table 1 shows the contents of refined iodized salt.

Table 1: Contents of Refined Iodized Salt

Sodium	≈39%
Chloride	≈60%
Ferrocyanide, Aluminum Silicate, Ammonium Citrate, Dextrose	Up to 2%
Iodide	.01%

Why is Salt Refined?

You may be asking yourself the above question. Salt is refined for four main reasons.

1. Refined salt, having all of its minerals removed (i.e., "purified") is essentially a lifeless product. Being a lifeless product

assures a long shelf life. In fact, refined salt can sit on the grocery shelf forever. A long shelf life is a valuable tool to maximize profits for food manufacturers.

2. Manufacturers believe that an all-white salt product will look cleaner to the consumer and, therefore, increase sales. Refined salt is bleached in order to obtain the white color.

3. If the salt is taken from a polluted area, the refining process will remove the toxins associated with the salt.

4. Iodine is added to refined salt to prevent goiter (swelling of the thyroid). However, as pointed out in my book, *Iodine, Why You Need It , Why You Can't Live Without It 4th Edition,* there is insufficient iodine in salt to prevent thyroid illnesses or to provide for the body's iodine needs.

Unrefined Salt

As contrasted with refined salt, unrefined salt contains much more than sodium and chloride. Unrefined salt contains all of the elements necessary for life. Celtic Sea Salt (Light Grey) contains 33% sodium, 50.9% chloride, 1.8% minerals and trace elements and 14.3 % moisture. Table 2 shows the major contents of unrefined Celtic Sea Salt.[2] Unrefined salt does not contain appreciable amounts of iodide.

Table 2: Major Contents of Unrefined Celtic Sea Salt

Element	%	Element	%
Chloride	50.9	Zinc	.00275
Sodium	33.00	Copper	.00195
Sulfur	0.820	Erbium	.00195
Magnesium	0.441	Tin	.00192
Potassium	0.227	Manganese	.0018
Calcium	0.128	Cerium	.00172
Silicon	0.052	Fluoride	.00109
Carbon	0.049	Rubidium	.00084
Iron	0.012	Gallium	.00083
Aluminum	0.0095	Boron	.00082
Praseodymium	0.0029	Titanium	00079
Strontium	0.00275	Bromine	.00071

How Is Unrefined Salt Harvested?

Unrefined salt has not been put through various machines to remove the minerals and other elements that are naturally part of the salt. In addition, unrefined salt has not been exposed to harsh chemicals. Unrefined salt should be collected without the use of heavy machinery or explosives. This minimizes the chance of contamination of the salt. Finally, unrefined salt should have the minerals and elements associated with its origin. These minerals will be apparent by the color of the unrefined salt.

Celtic Sea Salt

In the case of Celtic Sea Salt, the salt is harvested near the coast of northwestern France. Ocean water is channeled through a series of clay-lined ponds. Next, the wind and sun evaporate the ocean water, leaving mineral rich brine. When the salt farmer gathers the brine using wooden tools, salt crystals begin to form. This is essentially the same method of gathering salt as the ancient Celts used over 2,000 years ago.[3]

This gentle method of gathering the salt ensures that the salt will contain healthy minerals and other elements that are meant to be in salt.

There are other sources of unrefined salt besides Celtic Sea Salt. Redmond's Real Salt® is mined from salt deposits near Redmond, Utah. Redmond's salt does not undergo a refining process and therefore contains a wide array of minerals. Table 3 gives you the major contents of unrefined Redmond's Salt.[4]

Table 3: Unrefined Redmond's Salt

Element	%	Element	%
Chloride	59.1	Iodine	0.0009
Sodium	37.6	Manganese	0.0008
Calcium	0.418	Cesium	0.0007
Potassium	0.198	Erbium	0.00006
Rubidium	0.120	Phosphorus	0.00049
Sulfur	0.160	Titanium	0.00048
Magnesium	0.0937	Antimony	0.00042
Iron	0.0472	Cerium	0.00040
Silicon	0.0138	Zirconium	0.000389
Aluminum	0.0068	Barium	0.000291
Carbon	0.0060	Boron	0.000205
Silver	0.0030	Gadolinium	0.000199
Copper	0.0028	Samarium	0.000198
Bromine	0.0022	Strontium	0.000193
Fluoride	0.0013	Thallium	0.000133

Why You Should Use Unrefined Salt: The pH Factor

As previously mentioned, unrefined salt has many healthy minerals associated with it. On the other hand, refined salt contains primarily sodium and chloride as well as toxic additives. Unrefined salt is a whole food product which is easily utilized by the body. The additional minerals such as magnesium and potassium are essential for a healthy immune system. These additional agents are meant to be ingested at the same time as sodium and chloride are ingested.

Refined salt, in its highly processed form, is an unnatural substance to the body. Over millennium of time, our bodies were not exposed to salt as just sodium and chloride. Humans evolved over time using natural, unrefined salt with its full complement of minerals. Enzymes and hormones in our bodies were designed to utilize salt in its whole, natural form, not in a foreign, refined state. The consequence of utilizing salt in a devitalized form is a poorly functioning immune system, initiation and acceleration of chronic illness, and promotion of acidity.

Acidity and Alkalinity

The pH of the body is a measure of the acidity or alkalinity of the body. In an acidic body, the pH is lowered, while in an alkaline body, the pH is elevated. In a diseased state, the body can either too acidic or too alkaline. The body has many overlapping mechanisms designed to keep the pH of the body in a neutral state-around pH of 7.2.

When the pH of the body becomes either too acidic or too alkaline, normal physiologic functions decline. The organs of the body (kidneys, liver, brain, etc.,) do not function efficiently unless the pH of the body is neutral (≈7.2). The brain does not function well when there is either acidity or alkalinity in the body. Enzymes, the catalysts for the body, are very sensitive to pH changes. They will lose most of their function when the pH is altered. Enzymes can become deactivated with either an acidic or an alkaline pH. Immune system cells are unable to protect us when the pH is imbalanced. In fact, no part of the body will work efficiently if the pH is not properly balanced. An acidic pH is associated with many chronic illnesses including cancer, arthritis, osteoporosis, and Candida, as well as hormonal imbalances. Majid Ali, one of the foremost practitioners in holistic medicine has written that an acidic pH is a marker of the absence of health in the body. [5] An acidic pH is more common than an alkaline pH in a diseased state. Generally, the more ill one is, the more acidic their body is.

Food and pH

The food we eat can have a dramatic effect on pH. In nearly every case refined foods, devitalized of all the healthy vitamins, minerals, and enzymes, are acidifying to the body. Minerals are one of the most alkalinizing agents to the body. Due to poor diets full of refined foods, many people today are mineral deficient. Mineral deficiency is often associated with a lowered pH (<7.0). Refined salt has no minerals in it, while unrefined salt is loaded with minerals. It is not rocket science to realize that unrefined salt will be better to promote a healthy pH.

In fact, all refined foods, (including refined sugar, flour, oils, etc.) lack minerals, vitamins, and enzymes. When we eat these devitalized foods, our body has to use its own store of vitamins, minerals, and enzymes to break down food. Over time, this will lead to nutrient deficiencies and chronic illness. Furthermore, eating devitalized food leads to acidity in the body. Due the prevalence of refined food, it is no wonder that most people run on the acidic side.

My experience has shown that it is impossible to overcome chronic illness when there is an acidic condition present.

Cancer and chronic illness are two of the consequences of an acidic pH. Cancer cells will proliferate in an acidic environment. In fact, most chronic illnesses will occur in an acidic environment. It is very difficult to overcome any chronic illness if the pH of the body is acidic (i.e., pH<7.2).

In an experiment at home (performed by my daughter Jessi), one teaspoon of Celtic Sea Salt in ½ cup of filtered water increased the pH of the water from 6.4 (baseline) to 6.8-7.0. The same amount of refined salt, on the other hand, decreased the pH from 6.4 to 6.0. [6] Refined salt, lacking the buffering effect of the minerals, is an acidifying substance for the body. On the other hand, unrefined salt helps to maintain a more neutral pH and can actually help elevate an acidic pH.

In order to promote a more neutral pH, unrefined salt needs to be the salt of choice. Refined salt should be avoided at all costs.

Sue, 61 years old, has numerous food allergies. Over the years, she has become more and more allergic. Sue was reacting to nearly everything she was in contact with including foods. "I don't know what to eat. Everything seems to bother me," she complained. Now, she even has difficulty with taking supplements because she reacts to them. "I can't even take vitamins because they upset my body. I feel like I am allergic to everything," she said. Sue had tried different techniques to help her allergies with minimal effects. When I had Sue check the pH of her urine and saliva, she found that her pH was very acidic. Upon further investigation, Sue found that foods containing refined salt caused her pH to become more acidic. She said, "I could not believe that the foods that you eat can change the pH so dramatically." When Sue removed refined salt from her diet and added Celtic Sea Salt, her pH significantly increased. "The most important thing I found was that when my pH elevated, my food allergies went away. I also felt much better. My energy improved and I could think more clearly. Also, I am able to take supplements when my pH is elevated," she said. Now, Sue monitors her pH daily and adjusts her diet accordingly. Update on Sue: She is still monitoring her pH and using Celtic Sea Salt. She continues to do well If her pH stays above 6.5.

Final Thoughts

There is a huge difference between refined and unrefined salt. Unrefined salt is packed with essential minerals and supplies the body with a proper balance of sodium and chloride with over 80 trace minerals.

Refined salt is a poor food choice. It has no place in our diet. Without the balancing effect of the trace minerals, refined salt provides the body with too much sodium. Sodium was meant to be ingested with its complement of trace minerals. The consequences of the ingestion of large amounts of refined salt are mineral deficiencies, acidity, and the onset of chronic illness.

Unrefined salt should be the salt of choice. My clinical experience clearly shows that this is a healthier salt choice as compared to refined salt. Further evidence of this will be shown in the remaining chapters.

[1] Information on the refining process found at the Salt Institute:
www.saltinstitute.org

[2] From Grain and Salt Society

[3] From www.elementsofwellness.com/celtic salt.htm

[4] Western Analysis, Inc.

[5] Ali, M. "Of metalicized mouths, mycotoxicosis, and oxygen. Townsend Letter for Doctors and Patients. June 2005

[6] Thanks to Jessi Brownstein for performing this experiment.

Chapter 3

What Makes A Good Salt?
Minerals

What Makes a Good Salt? Minerals

Introduction

What makes a good salt? That is a key question to anyone interested in buying a healthy salt as well as anyone interested in the science behind salt.

In the last chapter, I showed the difference between unrefined and refined salt. Refined salt, which lacks minerals, is a devitalized, unnatural product that is best avoided. On the other hand, unrefined salt contains a full complement of minerals essential for maintaining optimal body function.

The most important factor in selecting a particular salt product is to choose one that contains minerals and has not been refined. The minerals in salt give the salt its particular color. Different salts have varied mineral levels depending on where they are harvested. Therefore, certain unrefined salts may look grey, while others may contain different colors including shades of red or green.

Refined salt products are not compatible with a healthy diet. You can usually recognize a highly refined salt product due to its pure white color and its very fine crystals. This was discussed in more detail in Chapter 2.

What About the Moisture?

As previously mentioned in Chapter 2, Celtic salt contains 14.3% moisture. This is contrasted with refined salt which contains no moisture. What is in this moisture and does it contain helpful nutrients for the body?

Many unrefined salts which come from the ocean contain moisture. This moisture forms during the processing and handling of the salt. Remember, most unrefined ocean salts are harvested from the oceans, through a process of evaporation. As the moisture evaporates, some of the moisture is absorbed into the salt crystals. To understand how this process works, we need to review the harvesting process as discussed in Chapter 2.

Most ocean salts are harvested through a process that allows some of the moisture of the brine to evaporate. Celtic sea salt is harvested by channeling the ocean water through a series of clay-lined ponds. The wind and sun are used to evaporate the sea water, leaving the mineral-rich brine. Eventually, the salt crystals will form from this mineral-rich substance.

Brine is water saturated or nearly saturated with salt. Brine is utilized in a variety of settings today. It is used as a preservative for vegetables, fish, and meat, as bacteria cannot grow in a brine solution. Before the advent of modern-day food storage with refrigerators, storing food with brine was done by nearly every family. In fact, the use of brine to store food dates back hundreds to thousands of years.

Unrefined Salt and Fluid Inclusions

As the brine is exposed to the air and sun, the moisture begins to evaporate from the brine. As the salt crystals begin to take shape, fluid inclusions form in the salt crystals. Fluid inclusions are small openings or cavities in the salt crystal. I envision these cavities as small caves in each salt crystal. As the moisture in the brine evaporates, some of the brine is left in these fluid inclusions.

An analysis of these fluid-filled inclusions found that they contain up to 35 times more magnesium and potassium as compared to sea water.[1]

I asked the manufacturer of Celtic Sea Salt to do an analysis of the brine on a batch of their Celtic Salt. The analysis was performed at a private lab (ENC Labs). The results are shown in Table 1.

Table 1: Comparison of Mineral Content of Celtic Salt and Brine

	Celtic Salt (%)	Brine (%)
Magnesium	0.5	3.19
Potassium	0.11	0.73

As can be seen from Table 1, the brine contains a much higher percentage of magnesium and potassium as compared to the actual salt.

The refining process of salt removes all the moisture from the final product. The manufacturers of refined salt attempt to remove all the moisture in order to allow the finished (i.e., refined) product to flow freely through salt shakers with very small openings. Although refined salt products will pour through small openings, they have had all of their valuable minerals removed.   This type of salt should be avoided.

Magnesium and Potassium

As can be seen from Table 1, the brine gathered from the ocean contains a high percentage of magnesium and potassium. These are two crucial minerals for the human body. Magnesium is the 11th most abundant compound in the human body. Magnesium is needed as a cofactor in hundreds of chemical reactions in the body. Adequate magnesium levels are essential for all living organisms. Unfortunately, most Americans are deficient in magnesium. Estimates are that only 32% of Americans receive the RDA for magnesium.[2] I routinely check red blood cell magnesium levels and find that over 75% of people are deficient in magnesium.

Potassium, like magnesium is essential for all living beings. It is a vital element for many important bodily functions including maintaining fluid balance, stimulating muscle contractions, and conducting nerve impulses. Most Americans are deficient in potassium.

Unrefined salt benefits the body by supplying it with a readily available substance that is rich in minerals such as potassium and magnesium. Unrefined salt is such a product.

Final Thoughts

Is it any wonder that most Americans are deficient in minerals? Ingesting refined food products, such as refined salt, is bound to lead to nutrient deficiencies.

Humans were designed to eat food to supply their bodies with the correct balance of nutrients in order for the body to function optimally. Unfortunately, the standard American diet will not supply these vital items; it will lead to a devitalized body lacking in basic nutrients.

Unrefined salt is a whole food source that supplies the body with a balance of over 80 essential minerals. I have no doubt that I would see less nutrient deficiencies, such as magnesium and potassium, if unrefined salt were the salt of choice for the vast majority of people.

---

[1] Geochimica et Cosmochimica Acta. Vol. 54. N. 14. July 1, 2001. 2293-2300
[2] USDA. http://www.ars.usda.gov/is/AR/archive/may04/energy0504.htm?pf=1.
Accessed 12.6.09

Chapter 4

Problems With Low-Salt Diets

Problems With Low-Salt Diets

Introduction

Low-salt diets have been recommended for many years. It is not too hard to find an article in a magazine or medical journal recommending that the readers lower their salt intake. Governmental agencies, the AMA, and many dietary groups have all recommended a low-salt diet. There are two questions to ask:

1. Is a low-salt diet helpful?
2. Is a low-salt diet healthy?

Is a Low Salt Diet Helpful?

The answer to this question depends on whom you are talking to and also depends on each individual's medical condition. For over 30 years, low-salt diets have been promoted by many organizations to help keep blood pressure low in the general population. Low-salt diets have also been recommended to lower blood pressure in hypertensive patients.

Does a low-salt diet lower the blood pressure? For the vast majority of people, the answer is no. This question will be explored in depth in this book. When looking at entire populations, the medical research is clear that a low-salt diet is not effective at significantly lowering the blood pressure. [1] [2] [3]

A low-salt diet can be helpful for certain individuals. People with hypertension have a better response to a low-salt diet than people without hypertension. However, the effect of salt restriction, even in those patients with hypertension, is modest, at best, with systolic blood pressures declining approximately 4.9 mm Hg and diastolic pressures declining 2.6mm Hg.[4] These numbers are not significant enough to cause the creation of public policy for promoting low-salt diets.

There is also a group of hypertensive patients who are salt sensitive. Salt sensitivity is defined as an increase in blood pressure due to a high sodium intake. Most hypertensive patients do not exhibit salt sensitivity. The only way to tell if an individual with hypertension will respond (via lowered blood pressure) to a low-salt diet is to institute a low-salt diet. The research shows that older individuals with hypertension will have a modest response. A review of 56 trials showed that a low-salt diet had minimal effect on blood pressure in the vast majority of people studied. In those studies, systolic blood pressure was lowered by an average of 3.7mm Hg and diastolic blood pressure was lowered by an average of 0.9mm Hg.[5] As previously mentioned, these numbers are nothing to set national policy with. This topic will be explored in more depth in Chapter 5.

Salt is excreted in the kidneys. Individuals with renal failure will have a decreased ability to clear salt from their diets. These individuals must watch their salt intake carefully. Because their kidneys are lacking the ability to clear excess salt from their diets, their bodies can become overloaded with salt. If you have renal failure, I suggest you speak with your doctor before instituting any dietary change, including a change in salt intake.

Is a Low-Salt Diet Healthy?

Everybody knows that a low-salt diet is healthy, right? Wrong. The research does not support that.

Low-salt diets are promoted by most conventional doctors, organizations like the American Heart Association, American Diabetes Association and many others. Though a low-salt diet may modestly improve the blood pressure in a salt-sensitive individual, a low-salt diet has other negative consequences in the body. This next section will deal with the possible negative consequences of a low-salt diet.

Low-Salt Diets and Myocardial Infarction (i.e., Heart Attacks)

As previously mentioned, low-sodium diets are widely recommended for cardiovascular benefit. It is thought that a low-sodium diet will result in lowered blood pressure and lowered risk of heart attacks.

Researchers studied the relationship between a low-sodium diet and cardiovascular mortality. Nearly 3,000 hypertensive subjects were studied. The result of this study was that there was a 430% increase in myocardial infarction (heart attack) in the group with the lowest salt intake versus the group with the highest salt intake.[6]

Why would a low-sodium diet predispose one to having a heart attack? Low-sodium diets have been shown to cause multiple nutrient deficiencies, including depletion of minerals such as calcium and magnesium, as well as depletion of potassium and B-vitamins.[7] There are numerous studies touting the benefits of magnesium in treating cardiovascular disorders.[8] [9] Adequate amounts of potassium and B-vitamins are also crucial for a healthy heart. Many studies have shown that a deficiency of minerals, particularly calcium, potassium, and magnesium, is directly related to the development of heart disease as well as hypertension. Furthermore, my experience has clearly shown that mineral deficiencies are present in most chronic disorders and it is impossible to overcome these disorders unless mineral deficits are corrected.

Low-Salt Diets, Cholesterol, and LDL Cholesterol

Higher cholesterol and LDL-cholesterol levels have been associated with adverse cardiovascular events including stroke and heart attacks. In fact, one of the most commonly prescribed class of drugs thought to decrease heart disease are the statins (e.g., Lipitor) which lower both total cholesterol and LDL cholesterol. Low-salt diets have been shown to cause significant increases (>10%) in both total cholesterol and LDL cholesterol.[10]

There is no question that diet impacts the risk of heart disease. Perhaps physicians should investigate dietary causes of high cholesterol/high LDL-cholesterol before resorting to statin drugs and their consequent side effects. My clinical experience has shown that unrefined salt, as part of a holistic program, has an ability to help lower both total cholesterol and LDL cholesterol levels. More information on statin drugs can be found in ***Drugs That Don't Work and Natural Therapies That Do, 2nd Edition.***

How About Other Hormones Influenced by Low-Salt Diets?

Maintaining adequate sodium levels is a crucial function of the body. Low-salt diets will result in the body increasing certain hormones (i.e., aldosterone, rennin, angiotensin and noradrenaline) to help the kidneys retain more sodium. These hormones can stimulate the sympathetic nervous system of the body. If the sympathetic nervous system is overly stressed, it can precipitate an adverse cardiac event such as a heart attack. Is it any wonder that a low-sodium diet may cause an increase in cardiovascular mortality?

In addition, the hormone, insulin, has also been shown to increase in a low-salt diet. [11] Elevated insulin levels have been associated with numerous metabolic disorders including diabetes, polycystic ovaries, and obesity. I have found it nearly impossible to treat insulin resistance and diabetes if the patient is on a low-salt diet. Additionally, the use of refined salt makes it nearly impossible to treat insulin resistance and diabetes. Unrefined salt is a necessity when treating any condition associated with elevated insulin levels.

Low-Salt Diets Promote Toxicity

A low-salt diet can lead to the accumulation of toxic elements in the body. The toxicity of bromide is exacerbated in a low-salt environment. Bromide is a toxic element that is associated with delirium, psychomotor retardation, schizophrenia, and hallucination.[12] Salt is a necessary ingredient in the diet that allows the body to detoxify toxic substances such as bromine. Chapter 9 will explore the toxicity of bromine in more detail.

Final Thoughts

Low-salt diets are promoted by conventional medicine as part of a healthy diet. However, low-salt diets are not associated with a reduction in blood pressure for the vast majority of the population. In addition, low-sodium diets have adverse effects on numerous metabolic markers including promoting elevated insulin levels and insulin resistance. Low-salt diets have been associated with elevating total cholesterol and LDL cholesterol levels which, in turn, have been associated with increases in cardiovascular events. Finally, low-salt diets will lead to mineral deficiencies and the development of chronic disease.

It seems clear that a low-salt diet is not only ineffective at controlling blood pressure, it is deleterious to the body. What conventional doctors and most mainstream organizations have failed to grasp is the difference between refined salt and unrefined salt. As mentioned previously, refined salt lacks minerals and causes acidosis (i.e., a lowered pH). Our bodies were meant to function optimally with adequate mineral levels and adequate salt intake. Only the use of unrefined salt and its' full complement of minerals can provide both of these factors.

[1] Smith, WCS, et al. Urinary electrolyte excretion, alcohol consumption, and blood pressure in the Scottish Heart Health Study. BMJ. 297:329-330, 1988

[2] Alderman, M., et al. Dietary sodium intake and mortality: The National Health and Nutrition Examination Survey (NHANES I). Lancet. Vol. 351, Issue 9105, March 14, 1998, 781-785

[3] Swales, JD. Salt saga continued: Salt has only small importance in hypertension. BMJ. 1988;297:307-8

[4] Cutler, JA., et al. An overview of randomized trials of sodium reduction and blood pressure. Hypertension. 1991;17 (suppl I): I-27-I-33

[5] Midgley, J., et al. Effect of reduced dietary sodium on blood pressure. JAMA. May 22/29, 1996. Vol. 275, No. 20

[6] Alderman, M. Low urinary sodium is associated with greater risk of myocardial infarction among treated hypertensive men. Hypertension, 1995;25:1144-1152

[7] Engstrom, A.M. et al. Nutritional consequences of reducing sodium intake. Ann. Intern. Med. 1983;98(part2):870-872

[8] Magnesium in treatment of acute myocardial infarction.
   Int J Cardiol. 2004 Sep;96(3):467-9.

[9] Magnesium intake and risk of coronary heart disease among men.
   J Am Coll Nutr. 2004 Feb;23(1):63-70.

[10] Ruppert, M . et al. Short term dietary sodium restriction increases serum lipids and insulin in Slat-sensitive and salt-resistant normotensive adults. Klin. Wochenschr. 1991;69: (suppl. XXV):51-57

[11] Rio, A. Del., et al. Metabolic effects of strict salt restriction in essential hypertensive patients. J. of Int. Med., 1993;233:409-414

[12] Levin, m. Bromide psychosis: four varieties. Am. J. Psych. 104:798-804, 1948

Chapter 5

Hypertension and Salt

Hypertension and Salt

Introduction

My medical training was clear: A low-salt diet was good and a high-salt diet was bad. I was taught to promote a low-salt diet in all hypertensive cases. In fact, I was taught that in order to prevent people from becoming hypertensive, it was better to encourage them to adopt a life-long dietary plan of low-salt. My experience with promoting a low-salt diet to treat hypertension was not successful. Not only did I find a low-salt diet relatively ineffective at lowering blood pressure, but I also found a low-salt diet made my patients miserable due to the poor taste of their low-salt food. Additionally, I rarely saw any positive benefits with a low-salt diet.

Early in my medical career, I accepted the "low salt=lowered blood pressure" hypothesis unquestionably. It wasn't until I began to look at my patients in a more holistic manner that I began to study the medical literature about salt. What I found was astounding; there is little data to support low-salt diets being effective at treating hypertension for the vast majority of people. Also, none of the studies looked at the use of unrefined salt, which contains many valuable vitamins and minerals such as magnesium and potassium, which are vital to maintaining normal blood pressure. This chapter will explore this topic in more detail.

Jack, age 63, had been treated for hypertension for seven years. He was taking two antihypertensive medications in order to control his blood pressure. He was on a diuretic (Dyazide) and a beta-blocker (Lopressor). "The pills work. My blood pressure is normal if I take the pills. However, I don't feel well on them. My energy level is gone and I am always tired. I can't have intercourse with my wife. Since I started the blood pressure medication I can't get an erection. Worst of all, I feel like my brain has left my head. I can't keep anything straight," he said. Jack's complaints about low energy, sexual problems and brain dysfunction are common when taking antihypertensive medications. When I examined Jack, I found him to be deficient in most minerals. Furthermore, he was deficient in salt. When I told

Jack that he needed to use unrefined salt in his diet, he was incredulous. He said, "My other doctors told me to lower my salt intake. I stopped using salt on all foods. I buy only low-salt or no-salt food. I was worried that the additional salt would make my blood pressure go higher." Blood work showed Jack low in sodium. I encouraged him to use unrefined salt in his diet. Salt and additional mineral supplementation had a positive effect on Jack. "My food tasted better and I felt better. After a week of the salt and supplements, my head began to clear," Jack claimed. Within two months, Jack was able to drop one of his medications (Dyazide) and decrease the other (Lopressor) medication in half.

Jack's story is very common in my practice. I don't feel that a patient with high blood pressure has an "antihypertensive medication deficiency" syndrome requiring prescription medications. My clinical experience has been clear: when nutrient imbalances are corrected, blood pressure will normalize itself. Elevated blood pressure is a sign of a problem in the body. Searching for and treating the underlying problems causing an elevation in blood pressure is the correct path to pursue.

Salt and Hypertension: The History

The first report of a relationship between salt and high blood pressure came about in 1904. Two researchers, Ambard and Beujard, reported that salt deprivation was associated with lowered blood pressure in hypertensive patients.[1] Over the next 50 years, various animal models were examined to support the hypothesis of salt causing high blood pressure. In almost all of these studies, huge amounts of salt (only in the form of refined salt—sodium chloride) were given to the animals to induce a significant hypertensive effect. The usual intake of salt was 10-20 times greater than the recommended dosages for these animals. Due to the high amounts of salt given to these animals, the correlation to a human population should have been suspect. Furthermore, these studies were not done with unrefined salt and its full complement of minerals.

However, the effect of eliminating refined salt on these overdosed rodents was the dramatic lowering of blood pressure. Medical researchers seized on these results and erroneously extrapolated them to a human population. Since that time, the "low salt=low blood pressure" dogma has been accepted as gospel. In fact, in 1979, the Surgeon General issued a report, based on the above studies, that claimed salt was the cause of high blood pressure and a low-salt diet was necessary to combat this.[2]

From that moment on, governmental agencies, researchers, medical schools, and dieticians became obsessed with the idea of lowering the national salt intake to improve hypertension in the general population. Shortly after the release of the Surgeon General's report, Arthur Hull Hayes, Jr., commissioner of the FDA proclaimed, "I look forward to the day when the American public will be as conscious of sodium intake as of calorie intake. Sodium reduction must remain a general health goal for our nation."[3] Without any substantial studies verifying that this idea was valid, the idea that lowering salt consumption would help improve hypertension was adopted by the medical community

INTERSALT Study: Setting National and Worldwide Policy

The most popular study cited to prove the "increased salt=elevated blood pressure" link was the INTERSALT Trial. This study looked at over 10,000 subjects aged 20-59 from 52 centers in 39 countries. The authors of the study looked at the relationship between electrolyte excretion (i.e., salt in the urine) and blood pressure. A higher salt intake results in a larger amount of salt excreted in the urine. Although there was a slight relationship between blood pressure and sodium excretion in INTERSALT, a "smoking gun" could not be found. This study showed a mild decrease in blood pressure (3-6mm Hg systolic and 0-3mm Hg diastolic) when there was a dramatic decrease in salt excretion.[4]

There were four population centers that had significantly lowered salt in their diets and also had significantly lowered blood pressure. These four areas were nonacculturated populations: Yanomamo and Xingu tribes in Brazil, and tribes in Kenya and Papua, New Guinea. People in these four areas had extremely low intakes of alcohol as well as very low body weights. The authors of the study point out that blood pressure does not rise with age in these four centers as it does in western countries.

The Yanomamo Indians in Brazil were cited as a group that had very low salt intake and low blood pressure. In addition, the Yanomamo Indians' blood pressure did not elevate with age. Researchers and academics seized on this as the "smoking gun" linking elevated salt with increased blood pressure. However, what is not well publicized is that Yanomamo Indians rarely live beyond 50 years.[5] In addition, the Yanomamo Indians don't have obesity in their population nor do they drink alcohol, which are two factors directly correlated with elevated blood pressure. Certainly, it is a long stretch to suggest national policy in the United States or any Western country should be based on the Yanomamo Indians study.

Body mass index (i.e., obesity) and heavy alcohol intake were found to correlate with elevated blood pressure in the INTERSALT study and in many additional studies. Perhaps the absence of obesity and alcohol intake were responsible for the very low blood pressures in these nonacculturated groups.

Trying to find the smoking gun linking salt intake to the development of hypertension was not found with INTERSALT. The results were underwhelming. In fact, the authors of the study wrote, "Across the other 48 centers, sodium was ...not {significantly related} to median blood pressure or prevalence of high blood pressure. Across the {other} 48 centers, there was no consistent association between salt intake and blood pressure changes."[6]

Other Studies of Salt and Hypertension

From 1966-2001, two authors summarized the findings of 57 trials of people (mostly Caucasians) with normal blood pressure. Low-sodium diets resulted in a very slight decline of systolic blood pressure by 1.27mm Hg and diastolic blood pressure by 0.54mm Hg as compared to a high-sodium diet.[7]

In eight trials of blacks, with normal or elevated blood pressure, low sodium intake reduced systolic blood pressure by 6.44mm Hg and diastolic blood pressure by 1.98mm Hg as compared to a high sodium intake. Interestingly, the authors found that there was a significant increase in cholesterol, LDL cholesterol, and triglycerides as well as the hormones rennin, aldosterone, and noradrenaline in the low-sodium diet as compared to the high-sodium diet.[8] These elevated hormones can cause an increase in cardiovascular events. The elevated hormones are the body's attempt to try to hold onto the little salt that is present in the diet.

A low-salt diet has been promoted as healthier for not only blood pressure but for cardiovascular events (i.e., stroke or heart attack). Eleven trials, which included follow-up from six months to seven years, were reviewed. Researchers found that there was no difference in deaths and cardiovascular events between the low-salt groups and the high-salt groups. Systolic and diastolic blood pressure declined in the low-salt group by only 1.1 and 0.6mm Hg. The authors of this review commented that the miniscule lowering of blood pressure with a low-salt diet did not result in any significant health benefit. They also stated, "It is also very hard to keep on a low-salt diet."[9]

NHANES (National Health and Nutrition Examination Survey) and Salt

Every ten years, the U.S. government does an analysis of thousands of its citizens looking at various markers of health. One such marker has been the mineral intake of the participants and the relationship to hypertension.

NHANES I (1971-1973) found that inadequate levels of minerals (potassium and calcium) were the best predictor of the presence of hypertension.[10] Further analysis found that adequate intakes of whole foods including fruits, vegetables, and dairy products were associated with the lowest blood pressure.

Higher dietary sodium levels were not associated with hypertension in NHANES I. Surprisingly, there was an inverse correlation with dietary sodium levels and blood pressure. In other words, low-sodium diets (i.e., low-salt diets) were associated with higher blood pressure as compared to a high-salt diet.[11]

NHANES III (1988-1994) and NHANES IV (2001-2002) found that lowered mineral intake, including magnesium, potassium and calcium were common in people with hypertension. As found in the previous NHANES I study, low intake of dietary sodium (i.e., low-salt diet) was also associated with high blood pressure in NHANES III and NHANES IV.[12]

Researchers concluded, "Our analysis confirms once again that inadequate mineral intake (calcium, potassium and magnesium) is the dietary pattern that is the best predictor of elevated blood pressure in persons at increased risk of cardiovascular disease."[13]

The CDC's own data over the last 30 years clearly shows little relationship between low-salt diets and hypertension. This data unequivocally shows that ensuring adequate mineral intake is much more important to maintaining a low blood pressure.

Barbara, age 53, went to her primary care physician for an annual exam and was told she had high blood pressure. "I was feeling a little tired, but I did not expect this. I was shocked when my blood pressure reading was 165/100. My doctor immediately wrote me a prescription for an antihypertensive medication and he told me I would be on this medication for the rest of my life," she said. Barbara, who works in my office, told me the story the next day. She was uncomfortable being told to take a medication for life. When I took a history from Barbara, I found that she was on a low-salt diet. Barbara said, "I avoided salt at all costs because I thought it caused hypertension." Barbara's sodium level was low (137mmol/L normal 141-145mmmol/L) on her laboratory work. I placed Barbara on unrefined sea salt (½ tsp. per day) as well as a regimen of vitamins and minerals. In addition, I advised her to clean up her diet and eliminate all refined foods. Within two weeks, her blood pressure started to decline and within two months her blood pressure was averaging 110/70. "When I started the salt and the supplements, I immediately felt better. I sleep better and I no longer get up during the night. Also, I find that I need less sleep at night," she claimed. Barbara's blood pressure remains lowered today and she is not taking any antihypertensive medications.

Update on Barbara. It has now been five years since Barbara has been adding unrefined salt to her diet. She is still not *taking antihypertensive medications and continues to maintain a lowered blood pressure.*

Are Low-Salt Diets Harmful?

Could low-salt diets actually cause more cardiovascular problems? In addition to the NHANES data reported above, researchers have similarly reported that as compared with a high-salt diet, a low-salt diet has been associated with a greater than 400% increase in risk of myocardial infarction (heart attack) in men.[14] MRFIT, a National Heart and Lung Blood Institute screened 361,662 men for a primary prevention trial looking at the effect of various interventions to lower mortality from heart disease. A very low-salt diet may provide minimal help lowering blood pressure in those that have hypertension. However, in people with normal blood pressure, there is little or no benefit from a blood pressure standpoint or a cardiovascular standpoint to justify instituting a low-salt diet.

Low-salt diets can result in a low sodium state (hyponatremia). In the vast majority of hypertensive patients, low-salt diets have never been shown to lower blood pressure consistently. In fact, only a small minority of patients (those that are salt sensitive) will exhibit a blood pressure lowering benefit of from a low-salt diet. However, it has never been shown that a low-salt diet will reduce mortality among hypertensive men. In one study, 2937 patients with hypertension were studied for their sodium intake. Men with the lowest sodium intake had a 430% increased risk of heart attack versus men with a higher sodium intake.[15]

Low Salt Diets and Heart Attacks: The Insulin Connection

How do you explain the higher risk of heart attack in the low-salt group? Low-salt diets have been shown to raise fasting insulin levels.[16] Insulin resistance is a widespread problem and is associated with diabetes and cardiovascular diseases. Also, LDL-cholesterol levels have been shown to be elevated in individuals consuming a low-salt diet.[17] Elevated LDL-cholesterol levels have also been shown to be related to the onset of cardiovascular disease. One hypertensive researcher commented that a low-salt diet "is meaningful therapy for only a fraction of the hypertensive population and might be deleterious for the majority."[18] Chapters 7 and 8 will provide more information about the relationship of low-salt diets and health.

What about High-Salt Diets and Mortality?

In Japan, the coastal Japanese population is estimated to have a salt intake double the average U.S. salt intake--≈300mmol sodium per day. The average life expectancy in Japan exceeds the life expectancy in the U.S. Clearly, there are other factors besides salt intake and mortality.

Does Unrefined Salt Lower Blood Pressure?

We have established that a low-salt diet is not very effective at significantly lowering blood pressure in most people. In fact, as salt levels have declined in this country over the last 50 years, there has been no trend toward lowered blood pressures in the population.

Could unrefined salt usage result in a significantly lowered blood pressure? Many minerals, including magnesium and potassium have a direct antihypertensive effect. As previously mentioned, The National Health and Nutritional Examination Survey (NHANES) is the U.S. government's method of collecting data on the nutritional status of Americans. In 1984, the first NHANES study revealed that a pattern of low mineral intake, specifically magnesium, potassium and calcium was directly associated with hypertension.[19] Repeated measurements over 20 years have confirmed the relationship between low mineral intake and elevated blood pressure.[20]

Magnesium is nature's muscle-relaxing mineral. For those with muscle spasms, magnesium intake is critical to stopping muscle cramping and pain. In the blood vessels of the body, the smooth muscle surrounding the blood vessels can become

71

constricted, elevating blood pressure. Magnesium has been shown to have a direct effect upon the relaxation capability of the smooth muscles of the blood vessels.[21] My experience has clearly shown that it is impossible to treat an elevated blood pressure without ensuring adequate magnesium intake.

Potassium deficiency has also been shown to cause elevated blood pressure in many studies.[22] [23] [24]

Researchers looked at using mineral salts in 20 elderly hypertensive subjects. The mineral salt contained lowered amounts of sodium and elevated amounts of magnesium and potassium. Six months after using this mineral salt, a decline of 11 mm Hg systolic and 15mm Hg diastolic blood pressure was observed in 45% of the people studied.[25] This decline in blood pressure exceeded the decline observed in most other studies when only sodium intake was lowered. There is no doubt that the larger decline in blood pressure in this study was due to the additional minerals (magnesium and potassium) found in the salt.

As previously stated, refined salt primarily contains sodium and chloride. Unrefined salt has a wide range of minerals including potassium and magnesium. Unrefined salt provides the body with the complex of nutrients that it needs to function optimally. The use of unrefined salt will not cause an elevated blood pressure. Furthermore, due to its abundance of minerals, it can actually help lower the blood pressure in hypertensive patients.

Sandra, 58 years old, watched her blood pressure gradually climb over the years from 120/70 to 140/90. Her internist had prescribed antihypertensive medications which Sandra could not tolerate. "I can't take many drugs. They make me tired and crabby. I told the doctor that I would rather live with the higher blood pressure than feel crummy all the time," she claimed. When I examined Sandra, she was found to be low in many minerals (magnesium, potassium, calcium, and lithium). In addition, she was salt deficient and hormonally imbalanced. Sandra was placed on a regimen of unrefined sea salt (Celtic Sea Salt), minerals, and natural hormones. She said, "I had totally removed salt from my diet five years ago because I thought it was a healthy thing to do." I explained to Sandra the importance of salt in the diet and encouraged her to add unrefined sea salt to her regimen. "Within two weeks, I felt better. I went from using no salt to using it all the time. My blood pressure actually went down. My energy shot up and I was back to my old self. Everyone who saw me asked me what I was doing since I looked so much better," she happily claimed. After correcting all of the above problems, Sandra's blood pressure normalized to 120/70. Sandra did not have an antihypertensive medication deficiency. She needed a holistic treatment plan that corrected her nutritional deficits.

Why Would There Be Conflicting Studies on Sodium and Hypertension?

The salt/hypertension argument is confusing. Some studies claim a benefit of reduced blood pressure when sodium is limited; others show no change. Overall, the literature, which claims that lowering salt intake has a positive effect on blood pressure is not convincing.

The studies are confusing because these studies have missed the most crucial aspect of this argument. Refined salt is very different than unrefined salt. Salt was never intended to be utilized in a devitalized refined state. Refined salt is a very toxic substance for our bodies. None of the studies looking at salt intake and hypertension use unrefined salt.

My experience has shown that unrefined salt, with its full complement of minerals, does not elevate blood pressure in the vast majority of patients. In patients with low blood pressure, the addition of unrefined salt appropriately elevates blood pressure. Conversely, in patients with elevated blood pressure, the elimination of refined salt and the addition of unrefined salt routinely lowers blood pressure.

How Much Salt Should a Hypertensive Patient Consume?

Researchers have looked at numerous studies to arrive at their recommendations for sodium intake. Hypertensive patients can improve blood pressure modestly by limiting their sodium intake to 3-7 grams (approximately 1.5-7 teaspoons) of salt per day.

Too much of anything can be a problem for the body. Salt, like any substance, should not be taken in excess. Since refined salt is a toxic substance for the body, there should be no refined salt ingested in anyone's diet.

However, as has been pointed out earlier in this book, there is a great difference between refined and unrefined salt. I recommend only the use of unrefined salt in one's diet. This will supply the body with over 80 minerals that are useful for maintaining the normal functioning of the body. My experience has shown that the use of unrefined sea salt has not resulted in elevated blood pressure in my patients. The amount of unrefined salt is directly related to the amount of water you consume. The more water you consume, the more unrefined salt you should ingest. I would recommend using ¼ tsp of unrefined salt for every quart of water ingested. The addition of small amounts of unrefined salt to food or cooking will not adversely affect blood pressure or other health parameters in someone with normal kidney function. For those with kidney problems, you must consult your physician on the appropriate salt intake.

Final Thoughts

My experience has clearly shown the fallacy of low-salt diets. For the great majority of people a low-salt diet does not work. People have a hard time maintaining these diets and the expected results are suboptimal. Patients do not feel well when sodium levels are lowered. Their energy level drops and they develop hormonal and immune system imbalances. Furthermore, other laboratory tests (lipid parameters) generally show worsened signs.

Physicians need a greater awareness of the difference between unrefined and refined salt. Refined salt needs to be avoided—it is a toxic, dangerous substance that fails to provide the body with little benefit. Unrefined salt should be the salt of choice.

[1] Ambard, L. Causes de L'hypertensin anerielle. Arch. Gen. De Med. 1904:1:520-33

[2] U.S. Dept. of Health, Education and Welfare. Healthy people: Surgeon General's report on health promotion and disease prevention. 1979

[3] Science. Vol. 216, 2 April 1982

[4] Samler, P. British Medical Journal. 1996. May 18;312(7041):1249-53

[5] Elijovich, F. The Argument Against. J. Clin. Hypert. 6(6):335-339, 2004. From: Medscape.com/viewarticle/480719_2

[6]

[7] Jurgens, G., et al. Effects of low sodium diet versus high sodium diet on blood pressure, rennin, aldosterone, catecholamines, cholesterols and triglyceride. The Cochrane Database of Systemic Reviews. 2004. Issue 1. Art. No: CD004022.DOI: 10.1002/14651858.CD004022.pub2

[8] Jugrens, G. IBID. 2004

[9] Hooper, L., et al. Advice to reduce dietary salt for prevention of cardiovascular disease. The Cochrane Database of Systemic Reviews. 2004, Issue 1. Art. No.:CD003556. DOI: 10.1002/1461858.CD003656.pub. 2

[10] NHANES I. CDC.

[11] Townsend, M. Low mineral intake is associated with high systolic blood pressure in the Third and Fourth National Health and Nutrition Examination Surveys. Am. J. of Hypertension. Vol. 18, 2. Feb. 2005, p. 261

[12] Townsend, M. IBID. 2005

[13] Townsend, M. IBID. 2005

[14] Alderman, M. Low urinary sodium is associated with greater risk of myocardial infarction among treated hypertensive men. Hypertension. 1995. Jun;25(6):1144-52

[15] Alderman, M. Low urinary sodium is associated with greater risk of myocardial infarction among treated hypertensive men. Hypertension. 1995;25:1144-1152

[16] Am. J. of Hypertension. 1991;4:410-415

[17].Am. J. of Hypertension. 1991;4:410-415

[18] Krakoff,L. Is reduction of dietary salt a treatment for hypertension? Am. J. of Hypertension. 1991: 481-482

[19] Townsend, MS. Low mineral intake is associated with high systolic blood pressure in the Third and Fourth National Health and Nutrition Examination Surveys: could we all be right? Am. J. Hyperten. 2005. Feb;18(2 Pt1): 261

[20] Townsend, M.S. IBID. 2005

[21] Rosanoff, A. Magnesium and Hypertension. Clin. Calcium. 2005. Feb;15(2):255

[22] Townsend, MS. Low mineral intake is associated with high systolic blood pressure in the Third and Fourth National Health and Nutrition Examination Surveys: could we all be right? Am. J. Hyperten. 2005. Feb;18(2 Pt1): 261

[23] Weglicki, W. Potassium, magnesium and electrolyte imbalance and complications in disease management. Clin. Exp. Hypertn. 2005. Jan;27(1):95

[24] Fortherby, MD. Long term potassium supplementation lowers blood pressure in elderly hypertensive subjects. Int. J. Clin. Pract. 1997;51: 219

[25] Katz, A. Effect of a mineral salt diet on 24-h blood pressure monitoring in elderly hypertensive patients. J. of Human Hypertension. (1999) 13, 777

Chapter 6

Salt and Water

Salt and Water

Introduction

In order to understand the benefits that good, unrefined salt has in the body, you must understand the importance of water. Water and salt are inexorably tied together. The human body contains approximately 70% water and the brain contains around 80% water. The human body also has around 250 grams of salt in the adult and 14 grams of salt in a baby. Water and salt are necessary for metabolism, detoxification, and transportation of nutrients as well as optimal functioning of the hormonal, nervous, and immune systems.

The Nervous System and Salt

The nervous system transmits impulses via electricity through the body.   A clear example of this is when you hit your "funny bone" on your elbow and you feel the shooting impulse down your hand.   That shooting impulse is a nerve (ulnar) that is transmitting the signal down the pathway to the hand.   These impulses are mediated by sodium ions.

Sodium regulates the electrical charges throughout the body.   If sodium levels are too low or too high, there will be abnormal electrical signals.   The brain is very sensitive to sodium changes.   In fact, one symptom of abnormal sodium levels can be the onset of a seizure disorder.

Seizure   disorders   are   increasingly   common   today. Although anti-seizure medications can be extremely helpful at controlling the frequency and severity of a seizure disorder, these medications do not treat the underlying cause of the illness.

The brain relies on sodium and water to transmit nerve impulses.   Other minerals are also involved in nerve conduction including magnesium, calcium and potassium.   Without adequate amounts of all of these substances, it is impossible for the brain to function normally.

As previously mentioned, sodium can be obtained from salt. However, refined salt, although supplying sodium and chloride in adequate amounts, is sorely lacking in over 80 minerals. In fact, the use of refined salt will cause and worsen a mineral deficit in the body. Mineral imbalances will increase the severity of any chronic disorder, including seizure disorders.

I have witnessed many of my patients with seizure disorders significantly improve their condition by eliminating refined salt and adding unrefined salt.

Jerry, 12 years old, suffered from recurrent seizures. He was on two anti-seizure medications which helped decrease the frequency of his seizures, but did not fully take them away. When I measured Jerry's electrolyte levels, his sodium was low at 138mmol/L (normal 141-145mmol/L). The only salt Jerry received was the refined salt in food. "I never salted Jerry's food because I thought it was healthier to go without it," his mother claimed. When I had Jerry's mother increase his salt intake with unrefined Celtic sea salt, his seizures dramatically improved. He was able to stop one seizure medication and decrease the second medication by 50%. "As soon as he started to take the Celtic Sea Salt, his seizures improved. In fact, his memory improved and he became more alert," said his mother.

Dehydration: A Common Problem Today

Water regulation is very important for maintaining optimal brain function. Water carries nutrients to the brain and allows the brain to flush out toxic chemicals.

Dehydration is a very common problem today. Most people not only drink inadequate amounts of water, they also drink other substances that pull water out of the body such as:

1. Caffeinated coffee

2. Caffeinated tea

3. Caffeinated soda

4. Juices high in sugar

The above substances not only do not provide water for the body, they contain substances (e.g., caffeine, sugar) that accelerate water loss in the body. A chronically dehydrated body will lead to a poorly functioning brain and a poorly functioning nervous system. I have seen many conditions in the body improve simply by correcting a water and salt deficit.

Lisa, 31 years old, suffered from severe migraine headaches for ten years. "I get at least three migraine headaches per month. They are debilitating. I can't eat and I can't sleep. My head hurts for two days after a migraine. I feel miserable," she said. When I saw Lisa, she exhibited many signs of dehydration. Her skin felt dehydrated, she had ridges in her nails, and her tongue had a very dry, cracked appearance to it. Lisa never salted her food as she thought it was the healthy thing to do. She said, "I would purposely avoid salt. I buy low-salt food items and I don't even own a salt shaker." Lisa's sodium level was low (serum level 139mmol/L(normal 141-145mmol/L) and her blood work revealed a dehydrated state. I instructed Lisa to increase her water intake to two liters per day and to use ½ tsp of Celtic Sea Salt per day. Within seven days, she had a dramatic improvement. "I could not believe the difference. My headaches just melted away. I can't believe water and salt have that much of an impact on my body. My headaches are totally gone. It was a miracle," she happily commented.

Lisa's story has been repeated many times over in my practice. Water and salt (unrefined salt) are the two most common substances in our body. It is impossible to achieve your optimal functioning when there are deficits of water and salt present. This chapter will explore this concept in more detail.

The Connection Between Salt and Water

Water is the primary nutrient in the body. Water not only generates energy, it carries nutrients to all the tissues of the body. Dr. F. Batmanghelidj, author of ***Your Body's Many Cries for Water***, writes about the two oceans in the body, named the extracellular and the intracellular oceans.

The extracellular ocean contains white blood cells, red blood cells, as well as vitamins and minerals. The intracellular ocean contains the interior organelles of the cells. These two oceans are in constant flux. When the cells of the body produce waste products, these waste products are released into the extracellular fluid for discharge by the kidneys. A lack of water or salt will block this process and lead to cellular decline and death.

Unrefined Salt Corrects Dehydration

Unrefined salt is the substance that nourishes both oceans. The sodium in unrefined salt primarily concentrates in the extracellular ocean while the potassium concentrates in the intracellular ocean. Potassium is necessary in the interior of the cells to hold onto water. Other minerals such as magnesium and calcium, both share in helping the two oceans communicate with each other.

If the body senses a salt deficit, the kidneys will have to work extremely hard to hold onto salt in the body. Prolonged salt deprivation will eventually lead to kidney problems. When the situation worsens, sodium deprivation will lead to toxic cells and early cell death.

The long-term ingestion of refined salt leads to cellular problems as well as the onset of chronic illness. Refined salt, lacking potassium and other minerals, will not nourish the intracellular ocean. The osmotic pressure of the refined sodium in the extracellular ocean will pull water from the interior of the cell, leaving the cell in a dehydrated condition. Drinking water won't help; the cell needs minerals. As the cell becomes more dehydrated, waste products begin to build up in the cell, causing acidosis (i.e., low pH—see Chapter 2 for more information). Eventually, cell death will be accelerated. The medical consequences of this include an increase in chronic illnesses such as cancer, autoimmune disorders, and arthritis, as well as accelerated aging. Only the use of unrefined salt, with its full complement of minerals as well as adequate water intake can reverse this condition.

Final Thoughts

The first step to any plan to optimize your health is to ensure adequate intake of healthy salt and water. I firmly believe that salt and water form the foundation for any healthy treatment plan. It is impossible to achieve your optimal health in a dehydrated and salt-deficient state.

All functions of the body, including the immune system, hormonal system, nervous system, and the cardiovascular system depend on adequate water and salt intake for optimal functioning.

Unrefined salt, containing over 80 minerals, is the perfect source of salt for the body. It provides a proper balance of nutrients that the body can use. My clinical experience has clearly shown that refined salt is not a healthful substance for the body and its continued use will lead to the onset of chronic illness.

Chronic illness is often associated with nervous system disorders including anxiety, brain fog, headaches, and seizures. Only with adequate intake of unrefined salt and water can these disorders be reversed. I have seen it occur over and over again in my practice.

Chapter 7

Salt and the Adrenal Glands

Salt and the Adrenal Glands

Introduction

The adrenal glands are small glands located on the upper pole of the kidneys, near the middle of the back. There are two main layers of the adrenal glands: the adrenal cortex and the adrenal medulla. The adrenal glands are known as the "fight or flight" glands. In a stressful situation, the adrenal glands will secrete a hormone (epinephrine) to prepare our bodies for action. In a resting state, the adrenal glands are responsible for maintaining adequate energy levels, blood sugar control, blood pressure control, muscle strength and much more via the secretion of other hormones including:

1. Aldosterone
2. Cortisol
3. DHEA
4. Estrogen
5. Pregnenolone
6. Progesterone
7. Testosterone

A salt-deficient diet and/or a mineral-deficient diet will lead to a cascade of events that starts with suboptimal adrenal function and eventually leads to adrenal exhaustion. Adrenal exhaustion is manifested by low adrenal hormone output on laboratory testing and clinical symptoms such as fatigue.

Adrenal exhaustion (or adrenal fatigue) is an epidemic problem in today's stressful world. The consequences of adrenal exhaustion include fatigue, a poorly functioning immune system, cancer, thyroid disorders, obesity, arthritis, fibromyalgia, chronic fatigue syndrome, autoimmune disorders, as well as many other chronic illnesses.

The adrenal glands are responsible for regulating salt absorption in the body. An adequate amount of healthy salt (i.e., unrefined salt) is vital for the adrenal glands to function optimally. An excess of refined salt will lead to a state of depleted minerals and, ultimately, adrenal exhaustion. Adrenal exhaustion is frequently associated with immune system disorders. I believe one of the reasons we are seeing such an increase in immune system disorders is due to the following two conditions:

1. The lack of unrefined salt in our diet
2. The excess use of refined salt in our diet

At 35 years old, Judy had suffered from fibromyalgia for five years. "Fibromyalgia has ruined my life. I can't exercise because if I do, I hurt more. I wake up tired and I go through the day in a fog. I feel like I am 80 years old," she said. When I first saw Judy, she was very pale and had black circles under her eyes. In addition, her tongue was thick and coated and she had very thin fingernails. These are all signs of mineral and salt deficiencies. Judy also had orthostatic changes in her blood pressure measurement—her blood pressure fell ten points from lying down to standing. The falling blood pressure, coupled with many of her symptoms, pointed to adrenal exhaustion. Judy was also found to be deficient in many of the adrenal hormones, salt, as well as a variety of vitamins and minerals. Judy was placed on a holistic treatment plan that included natural, bioidentical hormones (DHEA, pregnenolone, testosterone, progesterone, and natural hydrocortisone). Furthermore, Judy was treated with unrefined sea salt and a diet consisting of whole foods and adequate amounts of water. She immediately felt better. "I could tell within a few days that I was waking up. I went from feeling like an old, sick person to feeling great again in a few weeks."

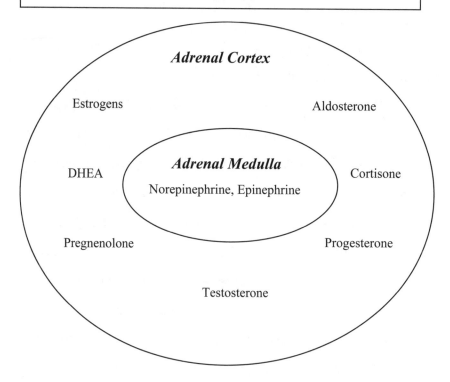

Figure 1: Adrenal Glands and the Hormones Produced

Physiology of the Adrenal Glands

In order to understand how salt affects the adrenal glands, you have to understand the physiology of the adrenal glands. The adrenal cortex is the outer portion of the adrenals. This area produces a number of hormones all aimed at regulating the basic vital functions of the body including blood sugar, blood pressure, and water and salt distribution, as well as muscle strength and energy. Figure 1 shows the different sections of the adrenal glands.

The hormones produced in the adrenal cortex include cortisone, DHEA, progesterone, the estrogens, testosterone, pregnenolone and aldosterone. All of these hormones have very important functions in the body. All of these hormones are stimulated from another pituitary hormone, ACTH (adrenal corticotrophic hormone). ACTH is produced in the brain from the pituitary gland.

ACTH also stimulates the middle layer of the adrenal gland, the adrenal medulla. One of the hormones secreted from this section of the adrenal glands is epinephrine. Epinephrine gives rise to the flight or fight response. It revs the body up. This "revving" of the body can also result in an anxiety state with panic attacks and elevated blood pressure if it is sustained for a long period of time.

Chronic stress on the body will result in adrenal exhaustion. Adrenal exhaustion occurs when the level of adrenal hormones (particularly the adrenal cortical hormones) decline. The symptoms one feels in a state of adrenal exhaustion include fatigue, low blood pressure, brain fog, inability to exercise, muscle aches and pains, joint pains, hair loss, thyroid problems, eczema and others. In addition, the thyroid gland will often malfunction when adrenal exhaustion is present.

The treatment of adrenal exhaustion requires many different modalities, including the use of natural hormones (refer to my book, **The Miracle of Natural Hormones 3rd Edition**) and thyroid hormones (refer to my book, **Overcoming Thyroid Disorders 2nd Edition**). In addition, adrenal exhaustion cannot be effectively treated without nutritional supplementation.

Adequate salt and mineral intake is essential for the adrenal glands to function optimally. Chronic overuse of refined salt will eventually lead to adrenal problems. Refined salt, lacking in minerals, does not supply the adrenal glands with the necessary items to perform at their most efficient level. Furthermore, the use of refined salt will not only be a causative factor in adrenal exhaustion; its continued use will inhibit the body from overcoming adrenal exhaustion. With refined salt being added to almost all processed foods, it is no wonder that adrenal exhaustion is occurring at such epidemic rates today.

Unrefined salt, on the other hand, not only contains easily absorbable forms of sodium and potassium; it also contains over 80 other minerals which are vital to many body functions. Magnesium is contained in unrefined salt in large amounts (and is absent in refined salt). Adequate magnesium intake is necessary for peak performance of the adrenal glands. Magnesium deficiency occurs in a large percentage of our population, partially due to the use of refined salt. Magnesium is nature's relaxing agent. It calms down irritated areas of the body and is an essential cofactor for the production of adrenal hormones.

Jackie, a 45 year old fitness instructor suffered from adrenal exhaustion for seven years. "I used to exercise every day. Now, I can't even train with my clients. If I exert myself too much, I pay for it for days. I feel like I have grown old," she exclaimed. Jackie had seen many doctors and had been diagnosed with fibromyalgia and chronic fatigue syndrome. Additionally, Jackie was plagued with palpitations. When I first saw Jackie, she had a very low blood pressure (90/55) and very tender muscles. When I asked Jackie if she used salt in her diet, her reply was "No way. I don't use salt for anything. I eat a healthy diet." I diagnosed Jackie with low mineral and electrolyte (sodium, potassium and chloride) levels, based on blood and hair testing. Upon placing her on a regimen of unrefined sea salt, minerals, and adrenal glandular support, she immediately felt better. "I felt like I was awakening from the dead. I slowly started to exercise and work out again. The salt was the biggest part. At first, I couldn't stand the taste of it, but now I really like it. In fact, I look forward to it," she said. Jackie made a full recovery from her illness over the next four months and now actively trains with her clients.

What about Cortisol?

The adrenal hormone cortisol regulates the immune system, vascular tone, and also helps with blood sugar regulation. Although cortisol is often vilified in advertisements (as causing obesity), adequate cortisol levels are necessary for normal energy production and a strong immune system. An acute stressful situation can initially result in elevated levels of cortisol. However, if the stress is prolonged, the adrenal glands will become exhausted. Adrenal exhaustion is often seen in laboratory tests as a depletion of cortisol (as well as other adrenal hormones). Depleted cortisol levels are associated with chronic illness including chronic fatigue syndrome, fibromyalgia, and autoimmune disorders. This deficiency can be easily discovered with a 24-hour urine test. My experience has shown that, in chronic illnesses, depressed cortisol levels occur in a large percentage of patients.

What about other Adrenal Hormones?

The adrenal hormones DHEA, progesterone, the estrogens, testosterone, and pregnenolone all have vital functions in the body including stimulating muscles, improving libido (in both men and women), and also improving brain function. In adrenal exhaustion, these hormone levels are very low. Supplementation with these hormones has shown marked benefits for patients suffering from adrenal exhaustion. For more information on these hormones I refer the reader to my book, **The Miracle of Natural Hormones, 3rd Edition.**

Aldosterone: An Important Adrenal Hormone

Aldosterone is referred to as a mineralcorticoid hormone since it helps regulate the minerals in the body. Aldosterone directly regulates the levels of sodium and potassium in the body. As previously stated in Chapter 6, there is fluid outside of the cells – the extracellular ocean – and there is fluid in the interior of the cells – the intracellular ocean. Sodium primarily concentrates on the outside of the cells of the body (in the outer ocean), while potassium is primarily found in the interior of the cell (in the intracellular ocean).

Sodium levels need to be properly maintained for hundreds of vital functions in the body. When there is a sodium deficiency present in the body, the adrenal glands respond to this lowered level by increasing their release of aldosterone. Aldosterone causes the kidneys to retain more sodium, therefore increasing sodium levels and reestablishing homeostasis. This is illustrated in Figure 2 (page 102).

How Do the Adrenal Glands Maintain a Normal Blood Pressure?

One of the main functions of the adrenal glands is to maintain a normal blood pressure. Good health cannot be maintained when the blood pressure is too high or too low. High blood pressure is associated with cardiovascular problems such as stroke and heart attack.

Low blood pressure is often found in those with fatigue states such as chronic fatigue syndrome or fibromyalgia. If the blood pressure is too low, an inability to exercise will be seen. In my office, I will often check my patients for their "orthostatic" blood pressure. I check their blood pressure when they are lying down and then when standing up. Normally, the blood pressure will increase >10mm Hg upon standing. The increase in blood pressure is necessary to get the body ready for more activity (in this case, standing up).

Adrenal exhaustion can lead to problems with maintaining a normal blood pressure. If the blood pressure declines from a supine to a standing position, it is a cardinal sign that the adrenal glands are not working correctly.

An elevated blood pressure is needed when we exercise or when we stand up in order to maintain adequate blood flow to the periphery of the body. When there is a decline in blood pressure with increased activity, it is termed "orthostatic hypotension". Orthostatic hypotension is very common with chronic fatigue syndrome, fibromyalgia and autoimmune disorders. Constant fatigue, dizziness, and feeling worse with exercise are the most common complaints I hear from patients with this condition.

Why is it so Important for the Adrenal Glands to Maintain a Normal Blood Pressure?

Without an adequate blood pressure, life is not possible. If the blood pressure falls too low, such as in times of trauma when there is severe blood loss, the organs of the body will not be perfused and death will result. Our body has many checks and balances to prevent this from happening. The adrenal glands are one of the main checks and balances responsible for preventing this.

When the body senses the blood pressure is too low, aldosterone secretion rises. Aldosterone causes the kidneys to retain more sodium from the diet. This results in fluid moving from the intracellular ocean to the extracellular ocean by osmosis. The end result is an elevated blood pressure. This is illustrated in Figure 2 (next page).

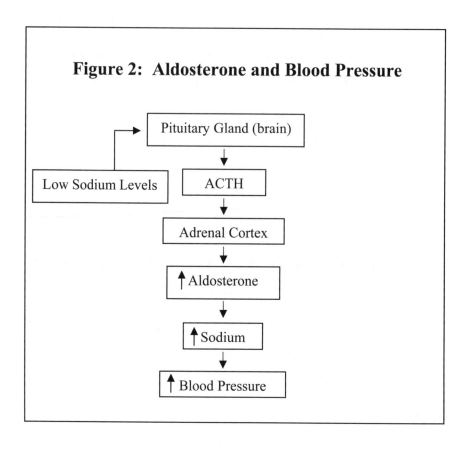

Figure 2: Aldosterone and Blood Pressure

Pituitary Gland (brain)

Low Sodium Levels

ACTH

Adrenal Cortex

↑ Aldosterone

↑ Sodium

↑ Blood Pressure

How Does a Low-Salt Diet Impact the Adrenal Glands?

A low-salt diet can actually make the body sodium deficient. The adrenal glands will sense an inadequate salt intake and take action to help the body absorb more salt from the diet. As depicted in Figure 2, aldosterone is the hormone secreted from the adrenal glands to help the body increase its absorption of sodium from the diet.

Sodium levels need to be maintained in a tight range in the serum. If not enough sodium is ingested, aldosterone levels will increase to help the body retain more sodium. This will result in water moving from the intracellular ocean to the extracellular ocean. If this situation occurs over a long time, the cells will become dehydrated and their energy production will decline. This can be the start of chronic illness.

Aldosterone also causes potassium and magnesium levels to fall. Ingesting refined salt will worsen the situation, since it is already deficient in magnesium and potassium. Thus not only will the cells be dehydrated, but they will also lack basic minerals such as magnesium and potassium. The end result is that the cells are unable to produce energy and fatigue will set in. Unhealthy cells are targets for cancer and other chronic illnesses to take hold.

Ben, age 41, got the flu five years ago. "One day I was well, the next day I had the flu. I expected to feel better within a few days. However, I never recovered. Now, I constantly feel tired and my brain is foggy. I can't even exercise, as it makes me more tired. As an attorney, I can't feel like this. I have to get my brain back," he said. Ben was diagnosed with chronic fatigue syndrome six months after he became ill with the flu. Ben said, "My doctor could make the diagnosis, but he could not offer a treatment plan. He said there is no treatment for chronic fatigue syndrome." When I saw Ben, he had many of the clinical signs of dehydration (dry skin, dry tongue, ridges on the nails, and dry skin). In addition, he had orthostatic hypotension, which is a sign that the adrenal glands are not optimally functioning. Before his laboratory results were back, Ben was placed on unrefined sea salt (1/2 teaspoon twice per day) and told to drink two liters of pure water per day. When I called him to discuss his laboratory values two weeks later, he was already doing significantly better. "I couldn't believe it. After three days of water and salt, my energy started returning. Within a week, I was not exhausted after work. I was amazed that just water and salt could do this," he claimed. Ben was prescribed a holistic treatment plan of vitamins, minerals and natural hormones, based on his laboratory values. Within five months of starting this holistic treatment plan, Ben made a full recovery.

What About Refined Salt and the Adrenals?

The ingestion of refined salt causes problems with the adrenals. Refined salt, lacking basic minerals, results in a mineral deficient condition. Refined salt contains large amounts of sodium. This large amount of sodium, coupled with a lack of minerals, results in an imbalance between sodium and the other minerals. The end result is adrenal exhaustion.

In conventional medicine, lowered amounts of salt (with no differentiation between refined and unrefined salt) are recommended to lower blood pressure, even though this therapy has never been shown to be very effective for the vast majority of people. The Standard American Diet will lead to many nutritional deficiencies including lowered mineral levels. As the mineral levels fall and the sodium/mineral balance worsens, blood pressure will begin to rise. When the low salt recommendations do not work, often the next step is diuretic therapy. Diuretics work by forcing the kidneys to lose sodium and, by osmosis, water. As the water and sodium content of the blood vessels fall, blood pressure will lower.

Diuretics can be an effective tool to lower blood pressure. Certainly in severe conditions such as congestive heart failure, diuretics can be life saving. I do not recommend anyone stopping any prescription medicine without consulting with your physician.

However, for the vast majority of patients that are prescribed a diuretic, the use of a diuretic only treats only the symptom of the illness (elevated blood pressure) and not the underlying cause. In this case, too much sodium pulls water out of the cells into the extracellular ocean resulting in edema and elevated blood pressure. A diuretic medication will treat the symptom (i.e., edema and elevated blood pressure) but not the underlying cause – an imbalance between sodium and minerals. Diuretics not only cause the body to lose sodium; they also cause the loss of many minerals, such as calcium, magnesium, and potassium. The long-term use of diuretic medications will result in a state of chronic dehydration and mineral deficiency, which can lead to chronic illness such as cancer, diabetes, heart disease, as well as other immune disorders.

Only with the use of unrefined salt and mineral supplementation can adrenal exhaustion be reversed. Remember, unrefined salt contains over 80 minerals in an easily absorbable form. Our bodies were designed to utilize and store minerals from unrefined salt.

My experience has clearly shown that the use of unrefined salt does not lead to elevated blood pressure. In fact, those with elevated blood pressure will often find their blood pressure lowered when their mineral imbalances are corrected by using unrefined salt.

Final Note on Diuretics

For those on diuretics, I do not recommend stopping them. If a diuretic medication is necessary, there are potassium-sparing diuretics that are generally more tolerated then other diuretic medications, as they do not promote a depletion of potassium. When possible, I suggest switching to one of the potassium-sparing diuretics. Consult with your physician before making any changes in your medications.

Final Thoughts

Aldosterone is the hormone produced by the adrenal glands which helps the body regulate blood pressure. In a low salt diet, aldosterone's main response is to tell the kidneys to retain more sodium for the body. More sodium in the body will cause the fluid level of the body to rise, resulting in an increased blood pressure and edema.

If the sodium primarily ingested comes from refined salt, there will be mineral deficits as well. The additional mineral deficits will worsen the condition resulting in an elevation of the blood pressure and edema. In this case, conventional antihypertensive therapies will be employed to treat only the symptoms of the illness – edema and elevated blood pressure.

Adequate amounts of unrefined sea salt, with its balance of minerals, supplies the adrenal glands with the support they need. This ensures that the adrenal glands will not need to increase the production of aldosterone and start the vicious cycle described in the chapter. Furthermore, adding unrefined salt to the diet can stop this cycle and lower the need for antihypertensive medications.

I have successfully treated hundreds of patients suffering from adrenal exhaustion with a holistic plan that includes the use of unrefined sea salt and increased water intake, as well as a regimen of vitamins and minerals. My clinical experience has clearly shown that an adequate amount of healthy salt (i.e., unrefined sea salt) is an essential ingredient for the maintenance of proper adrenal function as well as a healthy blood pressure.

Chapter 8

Salt and the Thyroid Gland

Salt and the Thyroid Gland

Introduction

Chapter 7 showed how refined salt and salt deficient diets can interfere with proper adrenal functioning. Adrenal stress will cause an imbalance in the adrenal hormones and will also cause an imbalance with other endocrine glands. All of the hormones of the body are related. It is like a symphony orchestra. If one endocrine gland is off balance, it will affect the other endocrine glands.

The adrenal and the thyroid glands are interrelated. If the adrenals are not functioning well, the thyroid will not function appropriately, and vice versa. This chapter will explore the thyroid gland and its relationship with salt intake.

Thyroid Hormone Production

The thyroid gland sits in the lower part of the neck and weighs approximately 1.5 ounces. It produces a teaspoon of thyroid hormone per year. This teaspoon of thyroid hormone has to run the metabolic machinery in every single cell in the body. Alterations in the amount of thyroid hormone produced, either too high or too low, will have large ramifications in the body. Hypothyroidism occurs when the level of thyroid hormone produced is suboptimal, while hyperthyroidism occurs when there is too much thyroid hormone produced. Figure 3 shows the pathway for thyroid hormone production.

Hypothyroidism

Hypothyroidism occurs when there is inadequate production of thyroid hormone. The symptoms of hypothyroidism are shown in shown in Table 4.

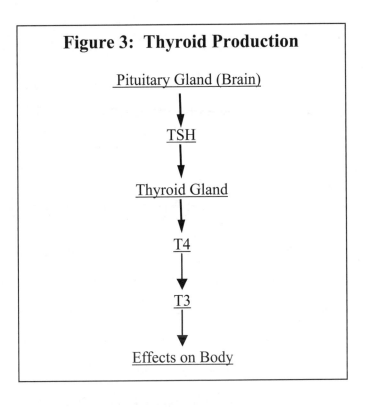

Figure 3: Thyroid Production

Pituitary Gland (Brain)

↓

TSH

↓

Thyroid Gland

↓

T4

↓

T3

↓

Effects on Body

Table 4: Symptoms of Hypothyroidism

Brittle Nails	Hypotension
Cold Hands and Feet	Inability to Concentrate
Cold Intolerance	Infertility
Constipation	Irritability
Depression	Menstrual Irregularities
Difficulty Swallowing	Muscle Cramps
Dry Skin	Muscle Weakness
Elevated Cholesterol	Nervousness
Essential Hypertension	Poor Memory
Eyelid Swelling	Puffy Eyes
Fatigue	Slower Heartbeat
Hair Loss	Throat Pain
Hoarseness	Weight Gain

Why Does Hypothyroidism Occur?

Hypothyroidism is a multi-factorial illness. Many different conditions can cause a hypothyroid condition including iodine deficiency, mineral deficiencies, heavy metal toxicities and genetic tendencies.

Why Does Hyperthyroidism Occur?

Hyperthyroidism is also a multi-factorial illness. Many of the same conditions that cause hypothyroidism can also cause hyperthyroidism. This includes iodine deficiency, mineral deficiencies, genetic tendencies, heavy metal toxicities, and infections. To find more information on thyroid disorders, I refer the reader to my book, ***Overcoming Thyroid Disorders, 2nd Edition.***

How Does Salt Impact Thyroid Disorders?

Salt impacts thyroid disorders on a number of fronts. First, there is the salt/iodine/thyroid connection. In the 19th century and early 20th century, the U.S. was plagued with an epidemic of swelling of the thyroid gland, or goiter.

Goiter was more common in areas where the soil was deficient in iodine. In the United States, this area was known as the Goiter Belt and it encompassed much of the central United States.

In the early 20th century, the Great Lakes States had the highest goiter rate of any area of the U.S. Due to the earlier work of researchers, it was hypothesized that adding iodine to the diet of people in the Great Lakes area would decrease the incidence of goiter. In 1923-1924, the State of Michigan's Department of Health conducted a large-scale survey of goiter in four counties. Of 66,000 school children examined, nearly 40% had enlargement of the thyroid gland (i.e., goiter).[1][2][3] In 1924, iodized salt was introduced to the area. By 1928, there was a 75% reduction of goiter observed, and by 1951, less than 0.5% of school-age children had a goiter. Research also showed a greater reduction in goiter among regular users of iodized salt as compared to non-users. Similar results were reported in Ohio from 1924-1950.[4]

Due to the positive results from using iodized salt in Michigan and Ohio, the rest of the United States quickly adopted the policy of adding iodine to salt, thus decreasing the goiter rate throughout the country. Today, the World Health Organization actively promotes the use of iodized salt to help prevent goiter throughout the world.

Iodized Salt and Thyroid Illnesses

The addition of iodine to salt has decreased the prevalence of goiter. However, as I describe in *__Iodine: Why You Need It, Why You Can't Live Without It, 4th Edition,__* the amount of iodine added to salt is not only inadequate for the thyroid gland, it is also inadequate for the body's iodine needs.

Thyroid illnesses, from hypothyroidism to autoimmune (hyperthyroid) disorders have been increasing at near epidemic rates over the last 30 years. During this time, studies by the National Health and Nutrition Examination Survey I (NHANES—1971-1974) and NHANES 2000 show iodine levels have dropped 50% in the United States.[5] This drop was seen in all demographic categories across the U.S.: ethnicity, region, economic status, population density, and race. The percentage of pregnant women with low iodine concentrations increased 690% over this time period. Low iodine concentrations in pregnant women have been shown to increase the risk for cretinism, mental retardation, attention deficit disorder, autism, and other health issues in the child.

Why Are Iodine Levels Falling When Iodized Salt Is Readily Available?

There are multiple reasons why iodine levels have fallen over the last 30 years. The stigma of salt causing high blood pressure has convinced many individuals to eat a low-salt diet. Without the proper supplementation of iodine, a low-salt diet will guarantee an iodine deficient state. Also, the addition of bromide (an inhibitor of iodine and a toxic agent) to our food supply has further worsened the condition. Furthermore, the toxicity of our environment with the exposure to fluoride and chlorine also inhibits iodine uptake in the body.

Also, the iodine added to refined salt is not bioavailable to the body. Researchers have estimated that only 10% of the iodine added to iodized salt is available for absorption.[6] This may be due in part to the bleaching agents (chlorine) used in salt that inhibit iodine absorption.

It is no wonder that iodine levels have been falling over the last 30 years. Iodine added to salt does help prevent goiter, but it does not prevent thyroid illnesses, nor does it provide nearly enough iodine for the rest of the body's needs. The consequences of iodine deficiency have been drastic, including increases in thyroid illnesses, cancer (including breast cancer), autoimmune disorders, and chronic illness.

Refined Salt and the Thyroid

The thyroid gland is very sensitive to the nutritional status of the body. If vitamins and minerals are deficient, the thyroid gland will not work appropriately. Thyroid hormone must be converted from its inactive (T4) form to its active (T3) form for the thyroid hormone to have beneficial effects on the body. Figure 4 illustrates what happens when there is a T4 conversion problem.

Thyroid hormone will not covert from an inactive (T4) to an active form (T3) resulting in many of the symptoms of hypothyroidism as detailed in Figure 4.

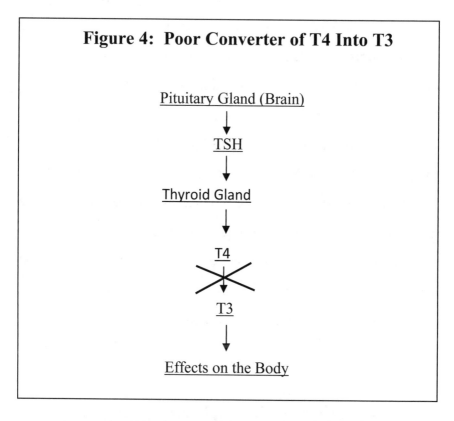

Items that block the T4 to T3 conversion include nutrient deficiencies. Selenium, magnesium, iodine, and other minerals impact the enzyme that facilitates the conversion of inactive (T4) to active (T3) thyroid hormone. Refined salt, lacking all of the basic minerals, will not provide the proper nutrition for the thyroid and its use will lead to a poor conversion of inactive (T4) to active (T3) thyroid hormone.

Refined salt, lacking minerals and containing toxic additives inhibits thyroid function and further leads to poor T4 to T3 conversion. On the other hand, unrefined salt, containing over 80 minerals, will help the thyroid function better and improve this conversion. My experience has clearly shown that unrefined salt nourishes the thyroid gland and promotes thyroid health. However, unrefined salt does not have iodine added to it. If you use unrefined salt in your diet, you may need to supplement with iodine.

Lori, a 56 year old nurse, had Hashimoto's thyroiditis for ten years. She was taking thyroid hormone medication (Armour thyroid). "The thyroid medication has definitely helped me. I can now think and I have the energy to get through my day. But, I have nothing left at the end of the day," she said. Lori also complained of swelling of her ankles and muscle cramps which sometimes woke her up at night. Lori avoided salt. "I thought it was healthier to never eat salt. That is what I hear every day at the hospital," she claimed. When I checked her laboratory tests, Lori was deficient in salt as well as iodine. Treatment with unrefined salt and iodine helped all of her remaining symptoms. "After introducing Celtic Sea Salt into my diet, I immediately felt an improvement. My leg cramps stopped and my skin even felt better. Now, after work, I am no longer tired and I don't have to put my feet up. The best part is, my food tastes better," she claimed.

Thyroid - Adrenal Connection

As stated in Chapter 7, the thyroid gland and the adrenal glands are closely related. Chapter 7 showed you that unrefined salt promotes proper adrenal functioning. If there is adrenal exhaustion present, the body cannot convert the thyroid from its inactive (T4) to its active (T3) form. Unrefined salt, with its ability to promote optimal adrenal function as well as proper thyroid function can help both glands function normally.

Final Thoughts

The use of refined salt will lead to many problems in the body, including hormonal problems. The thyroid gland is very sensitive to imbalances in the body. The constant ingestion of refined salt will lead to poorly functioning adrenal glands, thyroid disorders and the consequence of an imbalanced hormonal system.

Unrefined salt provides the proper balance of minerals for nourishing the endocrine system. These minerals are necessary for the thyroid and other endocrine glands to function optimally. For proper thyroid function, unrefined salt should be the salt of choice.

[1] Kimball, O.P. Prevention of goiter in Michigan and Ohio. JAMA. 1937; 108:860-864

[2] Matovinovic, J., et al. Goiter and other thyroid disease in Tecumseh, Michigan. JAMA. 1965: 192(#): 134-140

[3] Kimball, O.P. Endemic Goiter: A food deficiency disease. J. Am. Dietetic. Assn. 1949; 25:112

[4] Hamwi, G.J., et al. Endemic goiter in Ohio school children. Am.J. Pub. Health. 1955;45: 1344

[5] Hollowell, JE et al. Iodine nutrition in the United States. Trends and public health implications: Iodine excretion data from National Health and Nutrition Examination Surveys I and III (1971-74 and 1988-94). J Clin Endocrinol Metab 83:3401-3408. 1998.

[6] Abraham, G. The concept of orthoiodosupplementation and its clinical implications. The Original Internist 2004;12(1): 13-19

Chapter 9

Salt and Detoxification

Salt and Detoxification

Introduction

As stated in Chapter 4, salt can facilitate the detoxification of various harmful chemicals from the body. A low-salt diet will lessen the body's ability to detoxify many chemicals. In addition, a low-salt diet will exacerbate the toxicity of the element bromine.

Bromine: A Toxic Element

Bromine is a toxic element that is being ingested in increasing amounts in our modern world. More information on bromine can be found in ***Iodine: Why You Need It, Why You Can't Live Without It, 4th Edition.***

Bromine intoxication (i.e., bromism) has been shown to cause delirium, psychomotor retardation, schizophrenia, and hallucination.[1] Subjects who ingest enough bromine feel dull and apathetic and have difficulty concentrating.[2] Bromine can also cause severe depression, headaches, and irritability. It is unclear how much bromine must be absorbed before symptoms of bromism become apparent, but recent research has demonstrated that some symptoms of bromine toxicity can be present with low levels of bromine in the diet.[3] Bromine may also be implicated in cancer, particularly breast cancer. A study done at my office found that women with breast cancer had significantly higher levels of bromine as compared to women without breast cancer.[4]

Bromine (or its reduced form—bromide) is used as an antibacterial agent for pools and hot tubs. It is still used as a fumigant for agriculture. Crops sprayed with bromine have been found to have elevated bromine levels.[5] Bromine is also used as a fumigant for termites and other pests. In 1981, 6.3 million pounds of bromine were used in California. By 1991, that number increased to 18.7 million pounds.[6]   Toxicity of bromine has been reported from the ingestion of some carbonated drinks (e.g., Mountain Dew, AMP Energy Drink and some Gatorade products) which contain brominated vegetable oils.[7] Bromine is also found in most bakery products today, including breads, cookies, cakes, etc.[8]

Bromine in Medicine

Bromine used to be present in many common over-the-counter medications. It is still used today in many prescription medicines. Over 150 years ago, bromine was used extensively in medicine as a sedative as well as a remedy for seizures. Due to the toxicity of bromine, it has been phased out of many medicines.   However, bromine still can be found in some medicines including those that treat asthma, and bowel and bladder dysfunction (see Table 5 below). I believe all medicines that contain bromine need to be avoided.

Table 5: Currently Used Bromine-Containing Medications

Medicine	Indication
Atrovent Inhaler	Breathing Difficulties
Atrovent Nasal Spray	Breathing Difficulties
Ipratropium Nasal Spray	Breathing Difficulties
Pro-Panthine	Bladder Dysfunction
Pyridostigmine Bromine	Antidote for Nerve Gas

Bromine and Salt

As stated in Chapter 2, salt contains approximately 40% to 50% chloride. Chloride and bromine compete for reabsorption in the kidneys. When there is a decreased amount of chloride in the body, (which is common in low-salt diets), less bromine will be excreted from the kidneys resulting in elevated bromine levels. Increasing the amount of chloride in the diet will allow the kidneys to release more bromine into the urine for excretion.

When bromine is absorbed in the body, it tends to stay in the body for long periods of time. The "half-life" of a substance is referred to as the amount of time that 50% of the substance remains in the body after it has been absorbed into the body. A longer half-life would mean the substance would stay in the body for a longer time, versus a shorter half-life.

Studies show that one-half the dose of bromine is still in the human body 12 days after ingestion.[9] Therefore, the half-life of bromine is referred to as 12 days. In animals (rats), the half-life of bromine is 3 days. When rats are subjected to a low salt diet, the half-life of bromine is prolonged to 25 days—an 833% increase.[10] This means that a toxic substance—bromine—stays in the body for a longer period of time in a low-salt environment. The authors of this study conclude that a low-salt diet prolongs the elimination of bromine drastically.[11] Only with adequate chloride levels, can the body eliminate bromine in any significant amounts. Salt is the body's major source of chloride. In today's toxic world, our exposure to bromine is increasing. A low-salt diet will exacerbate the toxicity to bromine.

Bromine and Iodine

Bromine will bind in the body wherever iodine is bound. Iodine is an essential element for the production of thyroid hormone. When bromine binds to the thyroid gland, it is not only a toxic element but it can cause an iodine deficiency to occur. In conditions of iodine deficiency, bromine becomes more toxic.

Animal studies have shown that bromine intake can adversely affect the accumulation of iodide in the thyroid and the skin.[12] Research has also shown that a high bromine intake will result in iodide being eliminated from the thyroid gland and being replaced by bromine.[13] In addition, animal studies have shown that the ingestion of bromine can cause hypothyroidism.[14] When there is iodine deficiency present, the toxicity of bromine is accelerated. Therefore, maintaining adequate iodine levels is essential when you live in an environment that provides exposure to bromine.

HOW DO YOU LOWER BROMINE LEVELS?

Due to the addition of bromine in many food and drug sources (mentioned above), I believe that many of us have bromine toxicity, which is further exacerbated by iodine deficiency. In order to improve one's endocrine and immune system, a practical way of helping the body detoxify from bromine must be found.

There are simple ways of detoxifying from bromine as well as lowering the body's exposure to bromine. Primarily, we must stop ingesting bromine-containing food and medicines. That means eating organic food, grown without pesticides. Also, it means limiting bakery products that contain bromine.

However, once bromine is absorbed, it binds tightly to the iodine receptors in the body. Iodine supplementation allows the body to detoxify itself from bromine, while retaining iodine. Iodine works by competitively inhibit the binding of bromine which results in bromine being eliminated from the body.

Research has shown that bromine competes with iodide for absorption and uptake in the body.[15] Dr. Guy Abraham, a researcher on iodine, writes, "Therefore, increasing iodide intake should lower bromine levels in the thyroid preventing and reversing its thyrotoxic and goitrogenic effects."[16] The use of iodine will also cause bromine to be released from other tissues in the body as well as the thyroid. Dr. Abraham showed that increasing the intake of iodine would additionally increase the urinary excretion of other toxic halides.[17] In fact, Dr. Abraham has shown that iodine supplementation can result "in the whole body being detoxified" from the toxic elements bromine and fluoride. Research done in my office verified this statement from Dr. Abraham.

Putting It All Together: How to Decrease Bromine Levels with Salt

Although the use of iodine will displace bromine from its binding sites, the kidneys will not excrete bromine without adequate amounts of chloride. As previously mentioned, salt contains appreciable amounts of chloride.

My clinical experience has clearly shown that bromine will be excreted in larger amounts when the diet contains adequate amounts of salt. In fact, I have seen bromine excreted in large amounts in many of my patients when they are prescribed both iodine and unrefined salt. I have also found it nearly impossible to lower bromine levels when a low-salt diet is consumed.

Salt Bath

Soaking in a tub of warm water with 1-2 cups of unrefined salt (and two cups of hydrogen peroxide) is not only a soothing thing to do, it also facilitates the removal of toxins from the body. Salt baths will stimulate the lymph system to function better and will help excrete the toxins from the skin. The lymph system will also function much better when there is an adequate amount of salt present in the body.

My experience has shown that those patients with a compromised detoxification system in the body will benefit from salt baths. Salt baths have been effective at helping the body detoxify from heavy metals (mercury, lead, cadmium, etc.) as well as other toxic chemicals such as pesticides.

Final Thoughts

Low-salt diets will worsen the toxicity of bromine. An adequate salt intake will facilitate the excretion of the toxic element bromine from the kidneys. In fact, my experience has shown that low-salt diets will inhibit the detoxifying ability of the body for most toxic items.

It is nearly impossible to optimize the detoxification pathways in the liver when the patient is on a low-salt diet. Unrefined salt should be part of any healthy eating plan.

[1] Levin, M. Bromine psychosis: four varieties. Am. J. Psych. 104:798-804, 1948

[2] Clark. G. Applied Pharmacology. Churchill, London. 1938

[3] Sangster, B., et al. The influence of sodium bromine in man: A study in human volunteers with special emphasis on the endocrine and the central nervous system. Fd. Chem. Toxic., 21: 409-419, 1983

[4] Brownstein, D. To be published.

[5] Van Leeuwen, FX. The toxicology of bromine ion. Crit. Rev. Toxicol. 1987;18:189-213

[6] CAS Registry number: 74:83:9

[7] Horowitz, B. Bromism from excessive cola consumption. Clinical Toxicology, 35 (3), 315-320. 1997

[8] Abraham, G. The effect of ingestion of inorganic nonradioactive iodine/iodide in patients with simple goiter and in Graves' disease: A review of published studies compared with current trends. Optimox Reserch. 9.09.03

[9] Soremark, R. Excretion of bromine ions by human urine. Acta. Physiol. Scand. 50, 306. 1960

[10] Rauws, A. G. Pharmacokinetics of bromine ion-an overview. Chem. Toxic. Vol. 21, No. 1. 379. 1983

[11] Rauws. A.G. IBID. 1983

[12] Pavelka, S. High bromine intake affects the accumulation of iodide in the rat thyroid and skin. Biol. Trace elem. Res. 2001. summer;82(1-3):133

[13] Pavalka, S. Effect of high bromine levels in the organism on the biological half-life of iodine in the rat. Biol. Trace elem. Res. 2001. summer;82(1-3):133

[14] Buchberger, W. Effects of sodium bromine on the biosynthesis of thyroid hormones and brominated/iodinated thyronines. J. Trace Elem. Elec. Health Dis. Vol.4. 1990, p. 25-30

[15] vobecky, M. Effect of enhanced bromine intake on the concentration ratio I/Br in the rat thyroid gland. Bo. Trace. Element Res.: 1994

[16] Abraham, G. Iodine supplementation markedly increases urinary excretion of fluoride and bromine. Letter to the editor. Townsend Letter for Doctors and Patients. May 2003

[17] Abraham, G. IBID. 2003

Chapter 10

Uses of Salt

Uses of Salt

This Chapter will show how the use of unrefined sea salt can help with many common ailments.

Adrenal Exhaustion: Adrenal exhaustion is commonplace in our busy society. This topic was explored in depth in Chapter 7. Adequate salt intake is vitally important to restoring and maintaining optimal adrenal gland function.

Allergic Rhinitis (runny nose): Mix ¼ tsp of salt with ¼ tsp baking soda in 8 oz of pure water and use as a nasal spray. It will help lubricate and stop drainage of the nasal passages; it also acts as an antibacterial agent. In addition, salt has antihistamine properties.

Asthma: At the onset of wheezing, place one large pinch of salt on your tongue and drink 8 oz. of pure water (room temperature). Repeat in 15-30 minutes. If that doesn't work, use 1/4tsp of unrefined salt in 6oz of water. This regimen has helped me numerous times with my own asthma.

Circulation: Poor circulation is a common complaint. It often occurs in someone with low blood pressure. Salt helps to expand the blood vessel volume and can help improve circulation.

Detoxification: Adding 2 cups of salt and 2 cups of baking soda to the bath water can help stimulate the lymph system. This can aid in any detoxification plan.

Diabetes: Adequate salt intake is vital for diabetics. It is very difficult to control blood sugar in a salt deficient state.

Dry Skin: Rubbing your skin with salt after bathing or showering can help remove dry skin.

Exercise: At the start of exercise, take one large pinch of salt with a glass of water. If the exercise results in a large amount of sweating, repeat at the end of the exercise session. Salt will prevent muscle cramps and aid in muscle recovery after exercise.

Fatigue: Salt is necessary to produce energy in the body. Low salt diets will predispose one to fatigue.

Gastritis: Taking a large pinch of salt with meals helps to prevent reflux problems. Salt is an alkalinizing agent and helps to buffer excess acidity in the stomach.

Hyponatremia (low sodium levels): Low serum sodium levels are common both in those ingesting a very low-salt diet and in those on diuretics. Sodium is important for many different areas of the body including brain function, hormone production, and energy production. Those with low serum sodium levels (<141mmol/L) need to increase their salt intake.

Insect Bite including Bee Sting: Cover area with a warm salt paste. Pain and itching will be lessened.

Insomnia: One large pinch of salt with a small amount of warm water acts as a hypnotic agent.

Lipids: Low-salt diets will lead to poor lipid profiles. Chapter 3 explored this topic in more detail. Adequate salt intake will help improve lipid parameters.

Muscle Cramps: For nighttime cramps, take one large pinch of salt at bedtime with a small amount of water.

Osteoporosis: Salt is essential for proper bone density and strength. Osteoporosis is common with low salt diets.

Phlegm: Salt and water are the most effective expectorants known. One large pinch of salt and a glass of water will help to thin sputum.

Poison Ivy: Soak affected area in hot salt water (¼ tsp per quart of water).

Preservative: Salt has been used for thousands of years to help reserve food. There is no better preservative for canned foods than salt.

Chapter 11

Final Thoughts

Final Thoughts

I have no doubt that one of the main reasons I am seeing so many patients deficient in vital nutrients is due to poor food choices. The human body is designed to gather nutrients from food in order to maintain optimal body functions. These nutrients include vitamins, minerals, enzymes, and fatty acids.

Unfortunately, most of the new patients I see have mineral deficiencies. Minerals are necessary to catalyze thousands of reactions in the body. Minerals give the body its strength. For example, the bones contain the highest mineral content of any tissue in the body.

When I see a new patient in the office, I take a complete health history. Part of this history is taken by asking about their diet. Most patients think it is healthy to minimize salt in their diet.

In fact, if you are not eating unrefined salt in your diet, by my definition, you are salt deficient. The human body is designed to ingest salt. We have salt receptors on our tongue to give us the urge to eat salt. It is not only the tongue that has salt receptors. There are salt-regulating receptors in the kidneys and lining every cell in the body. All of the cells of the body require a certain makeup of nutrients (e.g., sodium, chloride, magnesium, and potassium) that are essential for maintaining optimal body functions. Why would the human body (and every other animal body) be designed with salt receptors that provide it with the urge to eat salt? We are designed that way because we cannot live without salt.

Do we get too much salt in our modern diet? Yes, we do get too much refined salt in our diet. In fact, any amount of refined salt is harmful for us. However, the vast majority of us do not ingest unrefined salt with its' full complement of minerals.

I wrote this book to educate the reader about the benefits of unrefined salt. I cannot emphasize these benefits enough. One of the reasons we are seeing so much illness in our modern world is due to the lack of nutrients in our diet. You can avoid becoming deficient in minerals by using unrefined salt in your diet.

To All Of Our Health!!

David Brownstein, M.D.

Resources

Selina Naturally™ is the owner of Celtic Sea Salt®
Brand.
For free recipes, articles, mineral analysis, or
ordering please call:

Selina Naturally™
4 Celtic Drive
Arden, NC 28704
1.800.-TOP-SALT (800.867.7258)
www.Celticseasalt.com

More information about Redmond's Real Salt
can be found at:
Redmond Trading Company, L.C.
475 West 910 South
Heber City, Utah 84302
1.800.367.7258
www.realsalt.com

Index

N

Noradrenaline 65

P

pH 30-34
Pesticides 132
Potassium 27, 29, 42-43, 67, 86, 143
Pro-Panthine 127

R

Redmond salt 29

S

Selenium 118
Seizure disorders 82-83
Sodium 82
Sulphuric acid 24

T

Triglycerides 65
Thyroid 112-120

Vytorin 69-70

Y

Yanomano Indians 64-65

Books by David Brownstein, M.D.

IODINE: WHY YOU NEED IT, WHY YOU CAN'T LIVE WITHOUT IT, 4th EDITION

Iodine is the most misunderstood nutrient. Dr. Brownstein shows you the benefit of supplementing with iodine. Iodine deficiency is rampant. Iodine deficiency is a world-wide problem and is at near epidemic levels in the United States. Most people wrongly assume that you get enough iodine from iodized salt. Dr. Brownstein convincingly shows you why it is vitally important to get your iodine levels measured. He shows you how iodine deficiency is related to:

- Breast cancer
- Hypothyroidism and Graves' disease
- Autoimmune illnesses
- Chronic Fatigue and Fibromyalgia
- Cancer of the prostate, ovaries and much more!

DRUGS THAT DON'T WORK and NATURAL THERAPIES THAT DO, 2nd Edition

Dr. Brownstein's newest book will show you why the most commonly prescribed drugs may not be your best choice. Dr. Brownstein shows why drugs have so many adverse effects. The following conditions are covered in this book: high cholesterol levels, depression, GERD and reflux esophagitis, osteoporosis, inflammation and hormone imbalances. He also gives examples of natural substances that can help the body heal.

See why the following drugs need to be avoided:

- Cholesterol-lowering drugs (statins such as Lipitor, Zocor, Mevacor, and Crestor and Zetia)
- Antidepressant drugs (SSRI's such as Prozac, Zoloft, Celexa, Paxil)
- Antacid drugs (H-2 blockers and PPI's such as Nexium, Prilosec, and Zantac)
- Osteoporosis drugs (Bisphosphonates such as Fosomax and Actonel, Zometa, and Boniva)
- Diabetes drugs (Metformin, Avandia, Glucotrol, etc.)
- Anti-inflammatory drugs (Celebrex, Vioxx, Motrin, Naprosyn, etc)
- Synthetic Hormones (Provera and Estrogen)

SALT YOUR WAY TO HEALTH, 2nd Edition

Dr. Brownstein dispels many of the myths of salt. Salt is bad for you. Salt causes hypertension. These are just a few of the myths Dr. Brownstein tackles in this book. He shows you how the right kind of salt--unrefined salt--can have a remarkable health benefit to the body. Refined salt is a toxic, devitalized substance for the body. Unrefined salt is a necessary ingredient for achieving your optimal health. See how adding unrefined salt to your diet can help you:

- Maintain a normal blood pressure
- Balance your hormones
- Optimize your immune system
- Lower your risk for heart disease
- Overcome chronic illness

THE MIRACLE OF NATURAL HORMONES, 3RD EDITION

Optimal health cannot be achieved with an imbalanced hormonal system. Dr. Brownstein's research on bioidentical hormones provides the reader with a plethora of information on the benefits of balancing the hormonal system with bioidentical, natural hormones. This book is in its third edition. This book gives actual case studies of the benefits of natural hormones.

See how balancing the hormonal system can help:

- Arthritis and autoimmune disorders
- Chronic fatigue syndrome and fibromyalgia
- Heart disease
- Hypothyroidism
- Menopausal symptoms
- And much more!

OVERCOMING THYROID DISORDERS, 2nd Edition

This book provides new insight into why thyroid disorders are frequently undiagnosed and how best to treat them. The holistic treatment plan outlined in this book will show you how safe and natural remedies can help improve your thyroid function and help you achieve your optimal health. NEW SECOND EDITION!

- Detoxification
- Diet
- Graves'
- Hashimoto's Disease
- Hypothyroidism
- And Much More!!

OVERCOMING ARTHRITIS

Dr. Brownstein shows you how a holistic approach can help you overcome arthritis, fibromyalgia, chromic fatigue syndrome, and other conditions. This approach encompasses the use of:

- Allergy elimination
- Detoxification
- Diet
- Natural, bioidentical hormones
- Vitamins and minerals
- Water

THE GUIDE TO HEALTHY EATING, 2nd Edition

Which food do you buy? Where to shop? How do you prepare food? This book will answer all of these questions and much more. Dr. Brownstein co-wrote this book with his nutritionist, Sheryl Shenefelt, C.N. Eating the healthiest way is the most important thing you can do. This book contains recipes and information on how best to feed your family. See how eating a healthier diet can help you:

- Avoid chronic illness
- Enhance your immune system
- Improve your family's nutrition

THE GUIDE TO A GLUTEN-FREE DIET, 2nd Edition

What would you say if 16% of the population (1/6) had a serious, life-threatening illness that was only being diagnosed correctly only 3% of the time? Gluten-sensitivity is the most frequently missed diagnosis in the U.S. This book will show how you can incorporate a healthier lifestyle by becoming gluten-free.

- Why you should become gluten-free
- What illnesses are associated with gluten sensitivity
- How to shop and cook gluten-free
- Where to find gluten-free resources

The Guide to a Dairy-free Diet

This book will show you why dairy is not a healthy food. Dr. Brownstein and Sheryl Shenefelt, CCN, will provide you the information you need to become dairy free. This book will dispel the myth that dairy from pasteurized milk is a healthy food choice. In fact, it is a devitalized food source which needs to be avoided.

Read this book to see why common dairy foods including milk cause:

- Osteoporosis
- Diabetes
- Allergies
- Asthma
- A Poor Immune System

Call 1-888-647-5616 or send a check or money order
BOOKS $15 each!

Sales Tax: For Michigan residents, please add $.90 per book.

Shipping :	1-3 Books	$5.00
	4-5 Books:	$4.00
	6-8 Books:	$3.00
	9 Books:	FREE SHIPPING!

VOLUME DISCOUNTS AVAILABLE. CALL 1-888-647-5616 FOR MORE
DVD's of Dr. Brownstein's Latest Lectures Available!
INFORMATION OR ORDER ON-LINE AT: WWW.DRBROWNSTEIN.COM

You can send a check to: Medical Alternatives Press
4173 Fieldbrook
West Bloomfield, MI 48323

DVD's of Dr. Brownstein's Newest Lectures
NOW Available!

All DVD's are from 1-2 hours. They are lectures Dr. Brownstein has given about the particular topic. Each lecture contains different information than what is in the books. Each DVD shows Dr. Brownstein lecturing and shows the slides he is lecturing about. More information about these products can be found at: www.drbrownstein.com

---

DVD: Why Use Holistic Medicine

Dr. Brownstein gives an overview of what is wrong with conventional medicine and which natural therapies can help you overcome illness. Dr. Brownstein will also provide you all new information on how to achieve your optimal health.

DVD: The Importance of Iodine

This lecture is a companion to Dr. Brownstein's newest book, <u>Iodine Why You Need It, Why You Can't Live Without It, 4th Edition.</u> Dr. Brownstein shows you why it is so important to ensure that you have adequate iodine levels. This lecture provides you with all the latest information on this important nutrient. Learn about:

DVD: Drugs That Don't Work and Natural Therapies That Do

This lecture covers Dr. Brownstein's latest book, <u>Drugs That Don't Work and Natural Therapies That Do.</u> Dr. Brownstein explains why the most commonly prescribed drugs should be avoided. Furthermore, he provides you with practical information about which natural therapies are effective and why they are much safer than the commonly prescribed prescription drugs. This lecture expands on the information contained in the book.

DVD: Overcoming Thyroid Disorders

A holistic approach to thyroid disorders is covered in this lecture. Dr. Brownstein reviews why so many people are suffering from thyroid disorders. He explains why the commonly used thyroid blood tests may not be the most accurate tests available. This book will provide you with how to optimize your thyroid and endocrine system.

DVD: Salt: Your Way to Health

Find out what the right kind of salt is and why you do need salt in your diet. Dr. Brownstein shows you why refined salt is a lifeless, devitalized product.

DVD: The Miracle of Natural Hormones

This lecture provides you with information about how safe and effective bioidentical, natural hormones are. Dr. Brownstein shows you why bioidentical hormones should be used in place of synthetic hormones. Actual case studies show the effectiveness of bioidentical hormones.

DVD's $25.00 Each (plus $5.00 s+h)
SPECIAL OFFER: Buy One Book for $15, Buy One DVD for $15!

(If ordering online, discount will be applied when order processed)

VOLUME DISCOUNTS AVAILABLE. CALL 1-888-647-5616 FOR MORE INFORMATION OR ORDER ON-LINE AT: <u>WWW.DRBROWNSTEIN.COM</u>

Call 1-888-647-5616 or send a check or money order to Medical Alternatives Press